Classical Vector Algebra

Every physicist and engineer, and certainly every mathematician, would undoubtedly agree that vector algebra is one of the basic mathematical instruments in their toolbox.

Classical Vector Algebra should be viewed as a prerequisite for, and an introduction to, other mathematical courses dealing with vectors, and it follows the typical form and appropriate rigor of more advanced mathematics texts.

The vector algebra discussed in this book briefly addresses vectors in general 3-dimensional Euclidean space, and then, in more detail, looks at vectors in Cartesian $\mathbf{R}^3$ space. These vectors are easier to visualize and their operational techniques are relatively simple, but they are necessary for the study of Vector Analysis. In addition, this book could serve as a good way to build up intuitive knowledge for more abstract structures of n-dimensional vector spaces.

Definitions, theorems, proofs, corollaries, examples, and so on are not useless formalism, even in an introductory treatise – they are the way mathematical thinking has to be structured. In other words, an "introduction" and "rigor" are not mutually exclusive.

The material in this book is neither difficult nor easy. The text is a serious exposition of a part of mathematics that students need to master in order to be proficient in the field. In addition to the detailed outline of the theory, the book contains literally hundreds of corresponding examples and exercises.

Textbooks in Mathematics
Series editors: Al Boggess, Kenneth H. Rosen

An Invitation to Abstract Algebra
Steven J. Rosenberg

Numerical Analysis and Scientific Computation (2nd Edition)
Jeffery J. Leader

Introduction to Linear Algebra: Computation, Application, and Theory
Mark J. DeBonis

The Elements of Advanced Mathematics, Fifth Edition
Steven G. Krantz

Differential Equations: Theory, Technique, and Practice, Third Edition
Steven G. Krantz

Real Analysis and Foundations, Fifth Edition
Steven G. Krantz

Geometry and Its Applications, Third Edition
Walter J. Meyer

Transition to Advanced Mathematics
Danilo R. Diedrichs and Stephen Lovett

Modeling Change and Uncertainty: Machine Learning and Other Techniques
William P. Fox and Robert E. Burks

Abstract Algebra: A First Course, Second Edition
Stephen Lovett

Multiplicative Differential Calculus
Svetlin Georgiev, Khaled Zennir

Applied Differential Equations: The Primary Course
Vladimir A. Dobrushkin

Mathematical Modeling the Life Sciences: Numerical Recipes in Python and MATLAB™
N. G. Cogan

Classical Analysis: An Approach through Problems
Hongwei Chen

Classical Vector Algebra
Vladimir Lepetic

Financial Mathematics Volume II: A Comprehensive Treatment in Continuous Time
Giuseppe Campolieti and Roman N. Makarov

www.routledge.com/Textbooks-in-Mathematics/book-series/CANDHTEXBOOMTH

Classical Vector Algebra

Vladimir Lepetic

CRC Press
Taylor & Francis Group
Boca Raton London New York

CRC Press is an imprint of the
Taylor & Francis Group, an **informa** business
A CHAPMAN & HALL BOOK

First edition published 2023
by CRC Press
6000 Broken Sound Parkway NW, Suite 300, Boca Raton, FL 33487-2742

and by CRC Press
4 Park Square, Milton Park, Abingdon, Oxon, OX14 4RN

CRC Press is an imprint of Taylor & Francis Group, LLC

ISBN: 9781032381008 (hbk)
ISBN: 9781032380995 (pbk)
ISBN: 9781003343486 (ebk)

DOI: 10.1201/9781003343486

Typeset in Palatino
by Newgen Publishing UK

Contents

Preface

Every physicist and engineer, and certainly every mathematician, would undoubtedly agree that vector algebra is among the basic mathematical instruments in their toolbox.

The title *Classical Vector Algebra* might be misconstrued as something particular, or something different from "simple" vector algebra. That is not the case. The adjective "classical", for lack of a better word, was used on purpose, for two reasons: first, in order to avoid the term "simple" which, arguably, is much disliked by students; second, to differentiate it from, say, Vector Calculus, Linear Algebra, or parts of Differential Geometry (which, of course, are separate fields on their own). In other words, the vector algebra discussed in this book briefly addresses vectors in general 3-dimensional Euclidean space, and then, in more detail, considers vectors in Cartesian $\mathbf{R}^3$ space. These vectors are easier to visualize, and their operational techniques are relatively simple, but they are necessary for the next step, which is the study of Vector Analysis. In addition, this book could serve as a good way to build up the intuition needed for more abstract structures of n-dimensional vector spaces.

Having said all that, the present book should be viewed as a prerequisite for, and an introduction to, other mathematical disciplines dealing with vectors, and it follows the typical form and appropriate rigor of more advanced math texts. Definitions, theorems, proofs, corollaries, examples, and so on are not useless formalism, even in an introductory treatise – they are the way in which mathematical thinking has to be structured. In other words, the terms "introduction" and "rigor" do not exclude one another – they should complement each other. This is not to say that the material in this book is difficult, nor that it is easy. It is simply an attempt to give a serious exposition of a part of mathematics that "everybody" working in the above-mentioned disciplines needs to master in order to be proficient in his/her field. In addition to the detailed outline of the theory, the book contains literally hundreds of corresponding examples and exercises. This author hopes that the reader will complete at least some of them.

Finally, the author would consider it a success if, after working carefully through this book, the reader is enticed to study more advanced mathematics.

The Author

Vladimir Lepetic is a professor in the Department of Mathematical Sciences, DePaul University. His research interests include mathematical physics, set theory, and the foundation and philosophy of mathematics.

1

Introduction

In natural sciences, particularly in physics, as well as in different engineering disciplines and certainly in mathematics, one distinguishes two basic quantities: ***scalars*** and ***vectors***. Unlike numbers that appeared almost "naturally" in all civilizations millennia ago, vectors, although ubiquitous, interestingly enough appeared in the sciences much later. In 1679 Gottfried Wilhelm Leibniz[1] apparently recognized the need to create an algebra capable of handling objects encapsulating both a magnitude and a direction. In 1687, in his *Principia*,[2] Isaac Newton, when discussing the problem of two forces acting on an object simultaneously, mentioned the diagonal of a parallelogram as the resultant sum of the acting forces. In the same period, other authors worked on the geometrical interpretation of complex numbers. Around 1830, Carl Friedrich Gauss,[3] following the work of Jean-Robert Argand,[4] published a paper describing entities comparable to complex numbers but placed in three-dimensional space. Works by W. R. Hamilton,[5] J. W. Gibbs,[6] O. Heaviside,[7] and H. Grassmann[8] followed. So, by 1910, vector analysis had become a standard tool of mathematicians and physicists.[9]

Notes

1 Gottfried Wilhelm Leibniz (1646–1716), German mathematician, philosopher, scientist, and diplomat.
2 Sir Isaac Newton (1643–1727), *Philosophiæ Naturalis Principia Mathematica*, 1687.
3 Johann Carl Friedrich Gauss (1777–1855), German mathematician.
4 Jean-Robert Argand (1768–1822), Swiss (amateur) mathematician.
5 Sir William Rowan Hamilton (1805–1865), Irish mathematician.
6 Josiah Willard Gibbs (1839–1903), American mathematician and physicist.
7 Oliver Heaviside (1850–1925), English mathematician and physicist.
8 Hermann Günther Grassmann (1809–1877), German mathematician, physicist, and linguist.
9 A reader interested in the history of vector analysis may consult Crowe, M.J., *A History of Vector Analysis*, Dover Publications; Revised ed. (November 2, 2011).

DOI: 10.1201/9781003343486-1

2

Vector Space – Definitions, Notation and Examples

As is always the case in science, we want to construct a mathematical formalism so that we can handle, with precision and rigor, the quantities that occur, in order to understand and describe natural phenomena and make predictions. The formalism we discuss in this book is (classical) **vector algebra.**[1] We will present it mostly through its geometrical model in intuitively understood 1-, 2-, and 3-dimensional spaces.

Indeed, since all natural physical phenomena occur in space, the first thing we ask is: what is a space? What is 1-dimensional, 2-dimensional, 3-dimensional, …, n-dimensional space? How should we think about it? How should we conceptualize it? The precise definition of a (vector) space will be given shortly. For the time being we approach it rather intuitively.

So, postponing formal definitions for the time being, we start with some frequently used "working" definitions. It is to be hoped that this will provoke curiosity, incite mathematical intuition, and prepare the beginner for the rigorous formalisms that follow.

Definition 2.1
Quantities completely specified by single data (real numbers), like length, mass, speed, temperature, and so on, are called *scalars*.

Quantities for whose accurate description we need two pieces of information, their magnitude and their direction, are called *vectors*. Typical examples of these are velocity, acceleration, force, electric field, and so on.

We assume that the reader is familiar with the concept of real numbers, i.e. the set **R**,[2] as well as the representation of real numbers on a (real) line. In other words, given a number $x \in \mathbf{R}$ we can "visualize" it as a point on a line (Figure 2.1).

The simplest example of a 1-dimensional space that we can think of would therefore be a real line, that is, the set **R**.

If we think of a line as a 1-dimensional space then we can think of 2-dimensional space – a plane – as a set of ordered pairs of real numbers:[3]

DOI: 10.1201/9781003343486-2

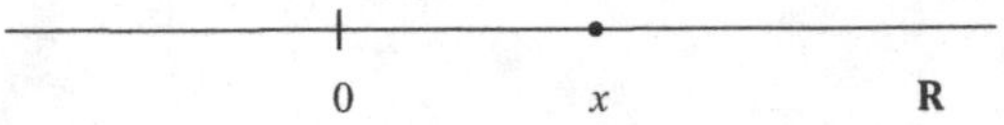

FIGURE 2.1

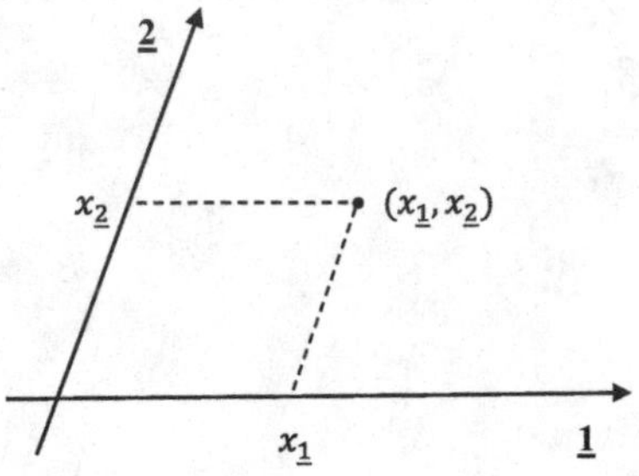

FIGURE 2.2

$$\mathbf{R}^2 = \{(x,y) \mid x, y \in \mathbf{R}\}$$
$$= \mathbf{R} \times \mathbf{R}$$

Thus, an ordered pair of two real numbers (x,y) represents a point in the plane, i.e. an element of a 2-dimensional space (Figure 2.2).

Similarly, a 3-dimensional space would be

$$\mathbf{R}^3 = \{(x,y,z) \mid x, y, z \in \mathbf{R}\}$$
$$= \mathbf{R}^2 \times \mathbf{R}$$

and a point in such a space, i.e. an element of this space, can be visualized as in Figure 2.3.

The fact that we cannot visualize spaces of higher dimensions does not mean that we cannot generalize the above concepts to higher dimensions, that is, to conceptualize an n-dimensional "point" $\boldsymbol{x}$ (a boldface x) and the space in which this point "lives". In order to do this, so that we are not restricted by the number of dimensions, we shall define a point in *n-dimensional space*, or, simply, *n-space*, to be an n-tuple of numbers

$$\boldsymbol{x} = (x_1, x_2, \dots, x_n)$$

where $n \in \mathbf{N}$.

Obviously,[4] the corresponding n-space is

$$\mathbf{R}^n = \{(x_1, x_2, \dots, x_n) \mid x_i \in \mathbf{R},\ i = 1, 2, \dots n\}.$$

We can think of the numbers $x_1, \dots, x_n$ as the coordinates of the "point" $\boldsymbol{x}$, and we can say that $\boldsymbol{x} \in \mathbf{R}^n$ is an element of the space $\mathbf{R}^n$.

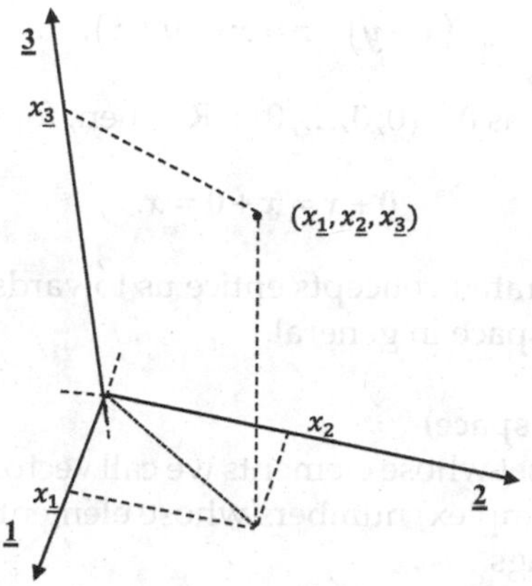

FIGURE 2.3

For $\boldsymbol{x}, \boldsymbol{y} \in \mathbf{R}^n$, we say that $\boldsymbol{x} = \boldsymbol{y}$, i.e. $(x_1, x_2, \ldots, x_n) = (y_1, y_2, \ldots, y_n)$, if and only if (iff) $x_1 = y_1,\ x_2 = y_2, \ldots, x_n = y_n$. Now we want to define the addition of those "points".

Definition 2.2
Let $\boldsymbol{x}, \boldsymbol{y} \in \mathbf{R}^n$. Then

$$\begin{aligned}\boldsymbol{x}+\boldsymbol{y} &= (x_1,\ x_2,\ \ldots,\ x_n)+(y_1,\ y_2,\ \ldots, y_n)\\ &= (x_1+y_1,\ x_2+y_2,\ \ldots,\ x_n+y_n).\end{aligned}$$

Since all $x_i\ , y_i \in \mathbf{R}$, we immediately see that the commutativity property of addition defined in this way holds. Namely,

$$\boldsymbol{x}+\boldsymbol{y} = \boldsymbol{y}+\boldsymbol{x}$$

The above short detour to one of the possible n-dimensional spaces is taken on purpose to indicate that many of the concepts of vector algebra discussed in this book can be generalized to more abstract structures and, vice versa, many of the concepts from n-dimensional spaces can be considered as "inherited" from classical vector algebra. The following example illustrates this.

Example 2.1
Let $\boldsymbol{x}, \boldsymbol{y} \in \mathbf{R}^3$, such that $\boldsymbol{x} = (1,2,3)$ and $\boldsymbol{y} = (3,1,2)$. Then

$$\begin{aligned}\boldsymbol{x}+\boldsymbol{y} &= (1, 2, 3)+(3, 1, 2)\\ &= (4,3,5)\end{aligned}$$

■

From Definition 2.2 and the example above, it is evident that, given $\boldsymbol{x}, \boldsymbol{y}, \boldsymbol{z} \in \mathbf{R}^n$

$$(x+y)+z=x+(y+z).$$

Now, if we define zero as $\mathbf{0}=(0,0,\ldots,0)\in\mathbf{R}^n$, then, for every $x\in\mathbf{R}^n$,

$$\mathbf{0}+x=x+\mathbf{0}=x.$$

These intuitively generated concepts entice us towards a formal definition of n-dimensional vector space in general.

Definition 2.3 (Vector space)
Let $X=\{x,y,\ldots,\}$ be a set whose elements we call vectors, and let $\Phi=\{\alpha,\beta,\ldots\}$ be the set of real (or complex) numbers whose elements we call scalars. Next, we define two mappings

$$\varphi:\Phi\times X\to X$$

and

$$f:X\times X\to X,$$

by

$$\varphi(\alpha x)=\alpha x,$$

$$f(x,y)=x+y,$$

such that, for every $\alpha,\beta\in\Phi$ and for every $x,y,z\in X$, the following axioms hold:

A.1 $\mathbf{x}+\mathbf{y}=y+x$;
A.2 $(\mathbf{x}+\mathbf{y})+z=x+(y+\mathbf{z})$;
A.3 There is a unique $\mathbf{0}\in X$, called the neutral element with respect to addition (additive identity), such that $\mathbf{0}+x=x+\mathbf{0}=x$;
A.4 $\exists(-\mathbf{x})\in X$, called the additive inverse, such that $\mathbf{x}+(-\mathbf{x})=(-\mathbf{x})+\mathbf{x}=\mathbf{0}$;
A.5 $\alpha(\mathbf{x}+\mathbf{y})=\alpha\mathbf{x}+\alpha\mathbf{y}$;
A.6 $(\alpha+\beta)\mathbf{x}=\alpha\mathbf{x}+\beta\mathbf{x}$;
A.7 $(\alpha\beta)\mathbf{x}=\alpha(\beta\mathbf{x})$;
A.8 $\exists 1\in\Phi$ such that $1\mathbf{x}=\mathbf{x}1=\mathbf{x},\forall\mathbf{x}\in X$.

We say that a quadruple $(X,\Phi;f,\varphi)=\mathscr{V}$ satisfying axioms A.1 to A.8 is a vector space (a linear space) over the field Φ[5]. Often, we simply say that X endowed with operations defined by the axioms A.1 to A.8 is a vector space.

The reader should note that the above definition is quite general, i.e. it says nothing about the dimension of the particular space. In other words, Definition 2.3 pertains to spaces of any dimension.

Notes

1 For the lack of a better term, we decided to call the vector algebra discussed in this book "classical" in order to distinguish algebra in 1-, 2-, or 3-dimensional spaces from the more abstract algebra of n-dimensional spaces.
2 A reader unfamiliar with the concept of a set and/or different sets of numbers should consult Appendices A and B.
3 A reader unfamiliar with set-theoretical symbols and notation should consult Appendix A.
4 Admittedly, "obvious(ly)" is an overused term in mathematical and scientific writing and can often be quite irritating. One needs to bear in mind that some mathematical statements that can be verified quickly might still not be obvious. (What is obvious for one person may not be obvious for another.) Calling something "obvious" means that the reason for its truth is or should be clearly understood. (cf. for example P. Halmos, *Linear Algebra Problem Book,* The Mathematical Association of America, 1995.)
5 The reader unfamiliar with the concept of a field should consult Appendix B.

3

Three-dimensional Vector Space $\mathscr{V}$

3.1 Definition and Basic Features of $\mathscr{V}$

Definition 3.1.1

Let's call our standard 3-dimensional Euclidean space $\boldsymbol{E}^3$ and (again, for the time being, without precisely defining this) let's assume that it has a point-like structure. Let $A(a_1, a_2, a_3)$ and $B(b_1, b_2, b_3)$ be two points in this space. We call an ordered pair of points (A, B) a vector $\overrightarrow{AB}$, where we distinguish between the beginning (initial) point, the "tail", A, and the end (final) point, the "tip", B. Equivalently, we say that a directed line segment $\overrightarrow{AB}$ for which we distinguish the beginning point A and the end point B is a vector. Also, with A and B defined in this way, we say that $\boldsymbol{a}_{AB} = \overrightarrow{AB}$ is a *displacement* from point A to point B. Following our intuition further, we represent a vector graphically as an arrow. In 2-dimensional space $\boldsymbol{E}^2$, a plane, we can visualize this as in Figure 3.1.

Analogously, we can visualize a vector in $\boldsymbol{E}^3$ as in Figure 3.2.

If $\boldsymbol{a} = \overrightarrow{AB}$ is a vector in $\mathbf{R}^2$, with $A(a_x, a_y)$ and $B(b_x, b_y)$, then its representation in the familiar Cartesian coordinate system would look like Figure 3.3.

Similarly, in $\mathbf{R}^3$ with $A(a_x, a_y, a_z)$ and $B(b_x, b_y, b_z)$ we would have what is shown in Figure 3.4:

Definition 3.1.2

Let S be the set of all oriented segments in $\boldsymbol{E}^3$. We say that the oriented segment $\overrightarrow{AB}$ is equivalent to the oriented segment $\overrightarrow{CD}$, and we write $\overrightarrow{AB} \equiv \overrightarrow{CD}$, iff they have the same length and the same orientation.

Or, equivalently,

Definition 3.1.2′

The oriented segment $\overrightarrow{AB}$ is equivalent to the oriented segment $\overrightarrow{CD}$, $(\overrightarrow{AB} \equiv \overrightarrow{CD})$, iff the segments $\overline{AD}$ and $\overline{BC}$ have a common midpoint (Figure 3.5).

DOI: 10.1201/9781003343486-3

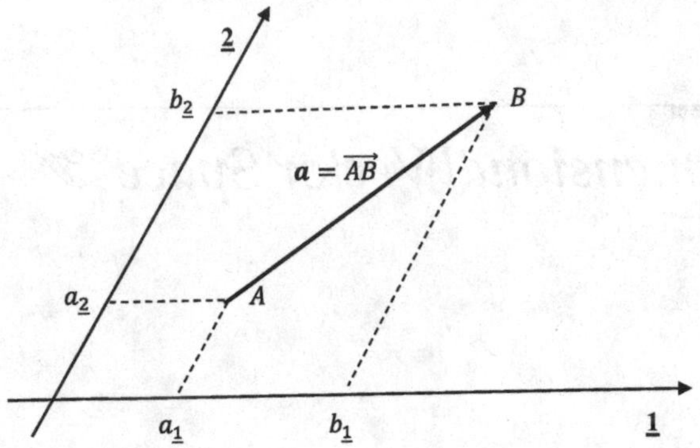

FIGURE 3.1

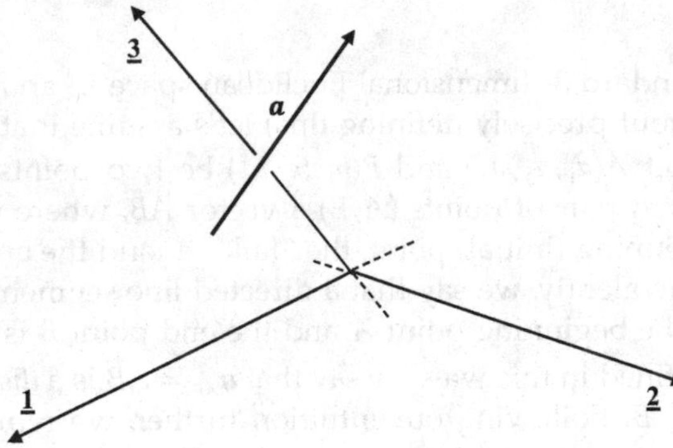

FIGURE 3.2

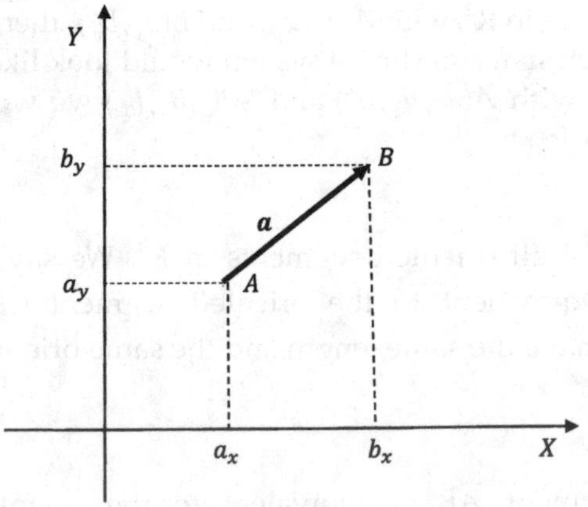

FIGURE 3.3

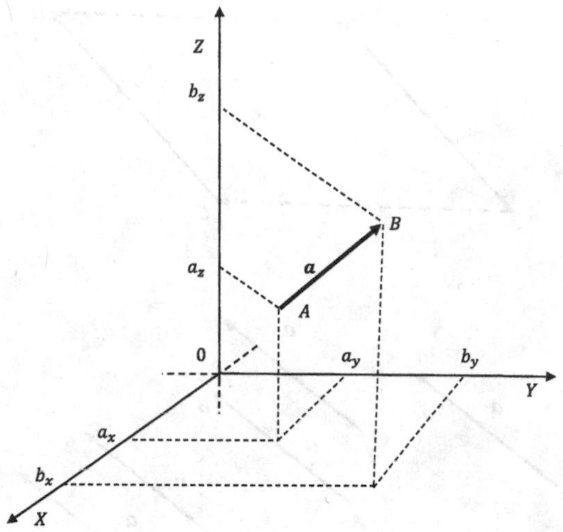

FIGURE 3.4

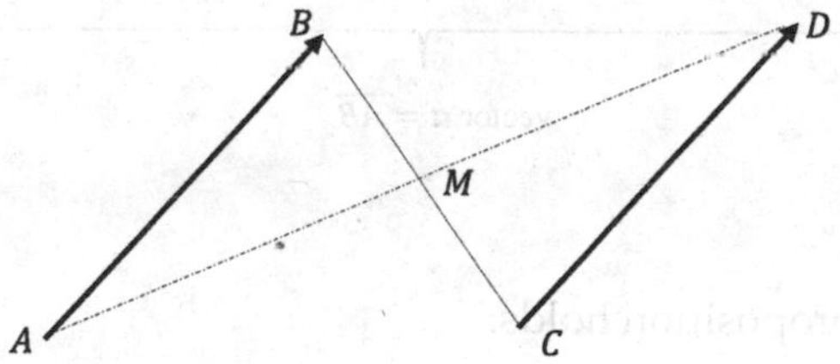

FIGURE 3.5

Example 3.1.1

If $\overrightarrow{AB} \equiv \overrightarrow{CD}$, then $\overrightarrow{AC} \equiv \overrightarrow{BD}$. Indeed, if $\overrightarrow{AB} \equiv \overrightarrow{CD}$, then $\overline{AD}$ and $\overline{BC}$ have a common midpoint, and therefore $\overrightarrow{AC} \equiv \overrightarrow{BD}$ (Figure 3.6). ■

Definition 3.1.3

Let S be the set of all oriented segments in $\boldsymbol{E}^3$, then we call the class $[\overrightarrow{AB}]$ of all oriented segments equivalent to $\overrightarrow{AB}$ a **vector**, and we designate it by $\vec{a}$ or a lower case **boldface** letter:

$$\boldsymbol{a} = \vec{a} = \left[\overrightarrow{AB}\right] = \left\{\overrightarrow{PQ} \mid \overrightarrow{PQ} \equiv \overrightarrow{AB}\right\}$$

From the definitions and the figures above, it is evident that $\overrightarrow{AB} \neq \overrightarrow{BA}$.

All this inspires us to say:

Every vector in ordinary 3-dimensional space is a ***representative*** *of an infinite family of vectors with the same magnitude*[1] *(length) and the same direction* (Figure 3.7).

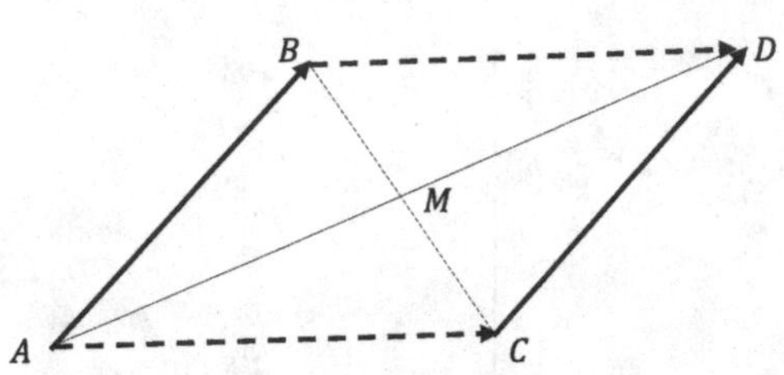

FIGURE 3.6

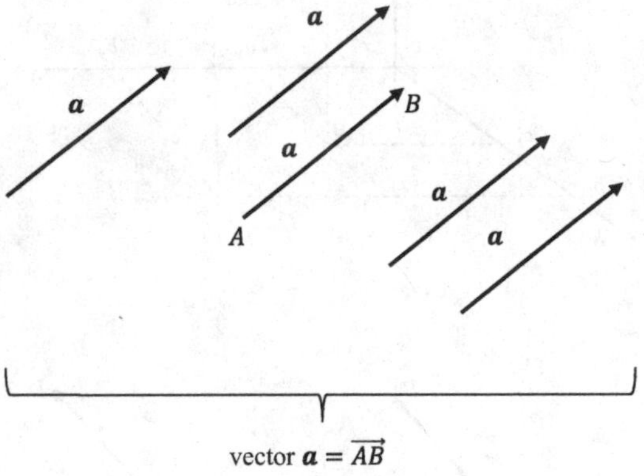

FIGURE 3.7

The following proposition holds:

Proposition 3.1.1
The relation "$\equiv$" on S is an equivalence relation. That is, it is:

(i) reflexive, i.e. $\overrightarrow{AB} \equiv \overrightarrow{AB}$, for all $\overrightarrow{AB} \in S$;
(ii) symmetric, i.e. if $\overrightarrow{AB} \equiv \overrightarrow{CD}$, then $\overrightarrow{CD} \equiv \overrightarrow{AB}$, for all $\overrightarrow{AB}, \overrightarrow{CD} \in S$;
(iii) transitive, i.e. if $\overrightarrow{AB} \equiv \overrightarrow{CD}$ and $\overrightarrow{CD} \equiv \overrightarrow{EF}$, then $\overrightarrow{AB} \equiv \overrightarrow{EF}$, for all $\overrightarrow{AB}, \overrightarrow{CD}, \overrightarrow{EF} \in S$.

So, to repeat, we will consider two vectors as "equal" iff they have the same magnitude and the same direction.

Definition 3.1.4
The set of all equivalent classes $S/\equiv$ of oriented segments obeying A.1 – A.8 in Definition 2.3 is called the (3-dimensional) vector space $\mathscr{V}$.

Proposition 3.1.2
Let $\overrightarrow{AB} \in S$ be any vector. Then for any point $C \in \boldsymbol{E}^3$ there exists a unique point $D \in \boldsymbol{E}^3$ such that $\overrightarrow{AB} \equiv \overrightarrow{CD}$.

Proof

Obviously, if $A = B$, then $D = C$. Suppose $A \neq B$, and let M be the midpoint of $\overline{BC}$, (Figure 3.6). Then choose a point D such that M is also the midpoint of $\overline{AD}$. By Definition 3.1.2',

$$\overrightarrow{AB} \equiv \overrightarrow{CD}.$$

■

Proposition 3.1.3

(i) $\overrightarrow{AB} \equiv \overrightarrow{CD}$ iff $\overrightarrow{AC} \equiv \overrightarrow{BD}$ iff $\overrightarrow{BA} \equiv \overrightarrow{DC}$;
(ii) If $\overrightarrow{AB} \equiv \overrightarrow{AC}$, then $B = C$;
(iii) If $\overrightarrow{AB} \equiv \overrightarrow{A'B'}$ and $\overrightarrow{BC} \equiv \overrightarrow{B'C'}$, then $\overrightarrow{AC} \equiv \overrightarrow{A'C'}$.

Proof

Let's prove (iii):

From (i) it follows that if $\overrightarrow{AB} \equiv \overrightarrow{A'B'}$, then $\overrightarrow{AA'} \equiv \overrightarrow{BB'}$. Also, if $\overrightarrow{BC} \equiv \overrightarrow{B'C'}$, then $\overrightarrow{BB'} \equiv \overrightarrow{CC'}$. Since $\equiv$ is equivalence relation, it is transitive, so it follows that $\overrightarrow{AA'} = \overrightarrow{CC'}$ and therefore $\overrightarrow{AC} \equiv \overrightarrow{A'C'}$.

Analogously one can prove (i) and (ii). ■

Proposition 3.1.4

Let $\boldsymbol{a} = \vec{a} \in \boldsymbol{E}^3$ be any vector, and let A be any point in $\boldsymbol{E}^3$. Then there exists a unique point $B \in \boldsymbol{E}^3$ such that $[\overrightarrow{AB}] = \boldsymbol{a}$.

Proof

Let $\boldsymbol{a} = [\overrightarrow{CD}]$. By Proposition 3.1.2 there exists a unique point $B \in \boldsymbol{E}^3$ such that $\overrightarrow{CD} = \overrightarrow{AB}$. Hence $\boldsymbol{a} = [\overrightarrow{AB}]$. ■

From now on, in order to simplify the writing and to be in notational agreement with the linear algebra of n-dimensional vector spaces, all vectors will be designated by boldface Latin letters, $\boldsymbol{a}, \boldsymbol{b}, \ldots, \boldsymbol{x}, \boldsymbol{y}, \boldsymbol{z}$, while scalars will be designated by regular print Greek letters $\alpha, \beta, \gamma, \ldots$.

Also, as will be explained in more detail later and to simplify operations with vectors, one often chooses the Cartesian coordinate system whose origin coincides with the tail of a given vector. To illustrate this let's do the following:

Consider the vector $\boldsymbol{a}$ shown in Figure 3.8 and place the origin of the coordinate system (not necessarily Cartesian) at the point A, or, equivalently, without changing the magnitude and the direction, move the vector $\mathbf{a} = \overrightarrow{AB}$ so that its tail coincides with the origin of the coordinate system. Let's call this vector $\mathbf{a}'$ (Figure 3.8).

By simple inspection we notice that the coordinates of the tail of the vector $\mathbf{a}'$ are $(0,0)$ and the coordinates of the tip, i.e. the point B', are $(\beta_1', \beta_2') = (\beta_1 - \alpha_1, \beta_2 - \alpha_2)$.

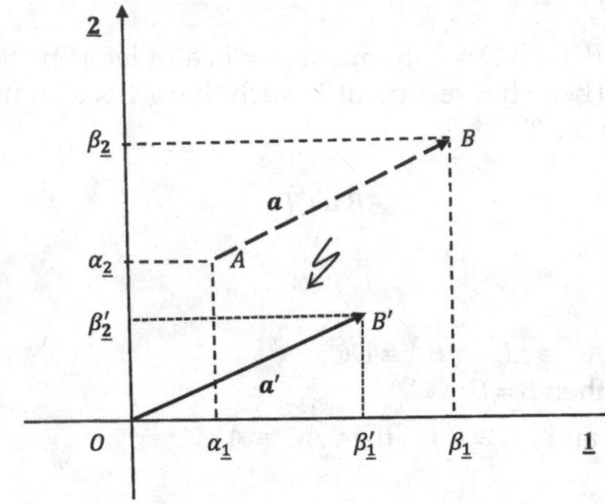

FIGURE 3.8

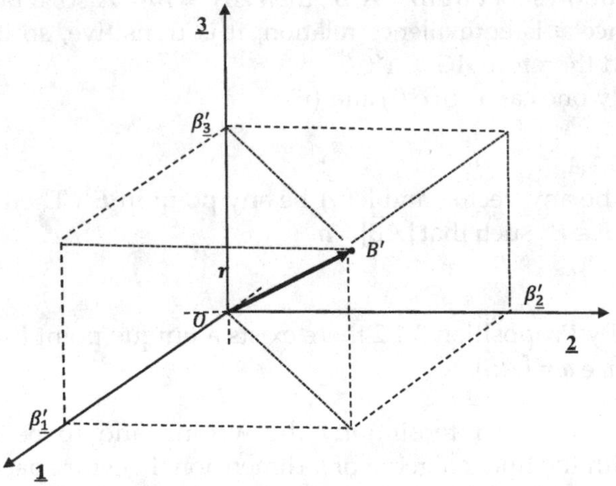

FIGURE 3.9

We say that vectors $\boldsymbol{a}$ and $\boldsymbol{a}'$ are *isomorphic*.

Finally, as the reader may find it more convincing, let's specify the points as, say, $A(2,3)$ and $B(5,6)$. Then the vector $\overrightarrow{AB} = \overrightarrow{(3,3)}$, as is the vector $\overrightarrow{O(B-A)} = \overrightarrow{(3,3)}$.

It is evident that (any) vector $\overrightarrow{OB'}$ with its tail located at the origin is completely determined by its end point. We will call this vector the **radius-vector** or **position vector** $\boldsymbol{r}$ (Figure 3.9).

Remark:

(i) It is customary that the components of a vector $\boldsymbol{r}$ in the Cartesian coordinate system are written simply as x, y and z, i.e. we write $\mathbf{r} = (x, y, z)$.

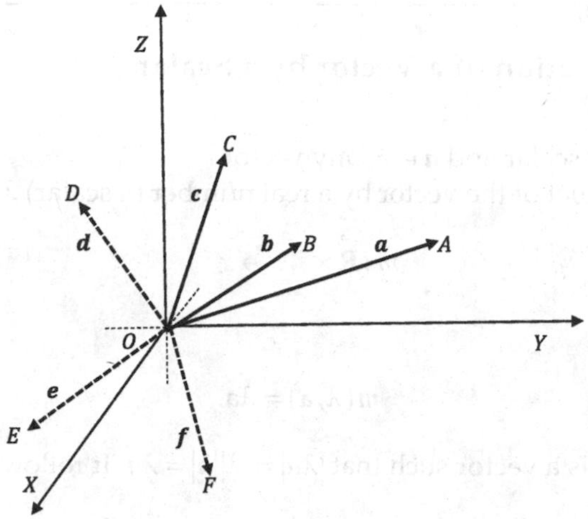

FIGURE 3.10

(ii) At this point the reader might already have anticipated that any vector, with adequate adjustments, can be represented as a radius vector. In other words, we can always move a vector in space so that its tail is at the origin while keeping its magnitude and orientation the same (cf. Definition 3.1.2 and the following propositions). So, the set of radius vectors, say, $a = \overrightarrow{OA}, b = \overrightarrow{OB}, c = \overrightarrow{OC}, d = \overrightarrow{OD}, e = \overrightarrow{OE}, f = \overrightarrow{OF}$ with $O(0,0,0)$ as their origin in $\mathscr{V}$ would look something like Figure 3.10:

Definition 3.1.5
Let $\boldsymbol{a} = [\overrightarrow{AB}] \in \mathscr{V}$ be any vector. We say that the *number* $a = |\boldsymbol{a}| = |\overrightarrow{AB}|$ representing the *length* of $\boldsymbol{a}$ is the **magnitude** or **modulus** of $\boldsymbol{a}$.

Definition 3.1.6
By the null-vector we mean the vector

$$\mathbf{0} = \vec{0} = \overrightarrow{AA}.$$

To be slightly more precise, and in accordance with Definition 3.1.3:

$$\mathbf{0} = \vec{0} = \{\overrightarrow{AA} \in \mathscr{V} \mid A \in E^3\}.$$

We consider the direction of the null-vector to be indeterminate, and, of course,

$$|\mathbf{0}| = 0.$$

3.2 Multiplication of a Vector by a Scalar

Definition 3.2.1
Let $\lambda \in \mathbf{R}$ be any scalar and $\boldsymbol{a} \in \mathscr{V}$ any vector.

Then the product of the vector by a real number (a scalar) λ is the mapping

$$m : \mathbf{R} \times \mathscr{V} \to \mathscr{V}$$

defined by

$$m(\lambda, \mathbf{a}) = \lambda \mathbf{a}.$$

The product $\lambda \boldsymbol{a}$ is a vector such that $|\lambda \boldsymbol{a}| = |\lambda||\boldsymbol{a}| = \lambda a$. It follows that:

(i) $1 \cdot \boldsymbol{a} = \boldsymbol{a}$;
(ii) $\lambda \boldsymbol{a} = \mathbf{0}$ iff $\lambda = 0$ or $\boldsymbol{a} = \mathbf{0}$;
(iii) $\lambda \boldsymbol{a} = -\boldsymbol{a}$ if $\lambda = -1$;
(iv) $\lambda \boldsymbol{a} = \boldsymbol{a} \lambda$;
(v) $\lambda(\mu \boldsymbol{a}) = (\lambda \mu) \boldsymbol{a}$;
(vi) $(\lambda + \mu) \boldsymbol{a} = \lambda \boldsymbol{a} + \mu \boldsymbol{a}$.

Thus, if $\boldsymbol{a} \neq \mathbf{0}$ and $\lambda > 0$, $\lambda \boldsymbol{a}$ has the same direction as $\boldsymbol{a}$. Of course, if $\boldsymbol{a} \neq \mathbf{0}$ and $\lambda < 0$, then $\lambda \boldsymbol{a}$ has the opposite direction to $\boldsymbol{a}$ (Figure 3.11).

Example 3.2.1
Let's prove (ii) from the above definition.

Take $\boldsymbol{a} \in \mathscr{V}$ to be any vector. Then, $\lambda \boldsymbol{a} = \mathbf{0}$ iff $\boldsymbol{a} = \mathbf{0}$ or $\lambda = 0$. Indeed,

$$\lambda \boldsymbol{a} = \mathbf{0} \Leftrightarrow |\lambda \boldsymbol{a}| = 0 \Leftrightarrow |\lambda||\boldsymbol{a}| = 0 \Leftrightarrow (|\lambda| = 0 \text{ or } |\boldsymbol{a}| = 0) \Leftrightarrow (\lambda = 0 \text{ or } \boldsymbol{a} = \mathbf{0}). \qquad \blacksquare$$

Exercise 3.2.1
Prove (iii) – (vi) above.

3.3 Collinear and Coplanar Vectors

Definition 3.3.1
Two, nontrivial, vectors $\boldsymbol{a}$ and $\boldsymbol{b}$ (i.e. $\boldsymbol{a} \neq \mathbf{0}, \boldsymbol{b} \neq \mathbf{0}$) are said to be ***collinear*** if there exists a number (a scalar) $\lambda \in \mathbf{R}$ such that $\boldsymbol{a} = \lambda \boldsymbol{b}$ (Figure 3.11). In other words, $\boldsymbol{a}$ and $\boldsymbol{b}$ are collinear when they are parallel to the same line. Analogously, one can define the collinearity of n vectors.

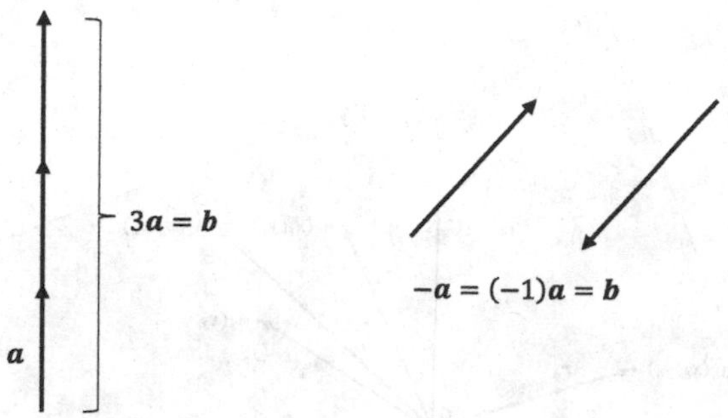

FIGURE 3.11

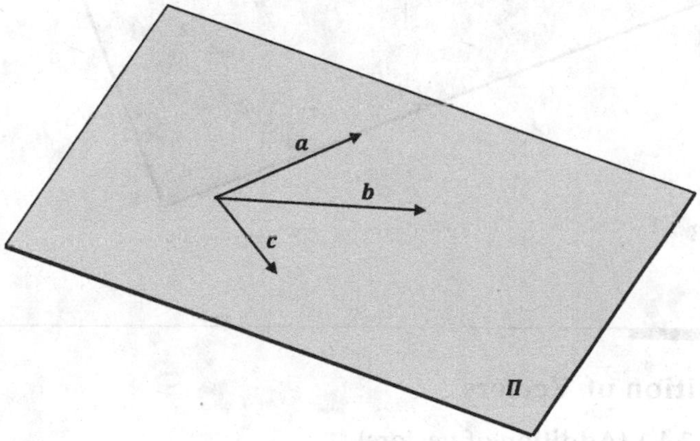

FIGURE 3.12

We will accept the convention that the zero vector (null-vector) is collinear with any vector.

We say that two vectors $\boldsymbol{a} = \overrightarrow{OA}$ and $\boldsymbol{b} = \overrightarrow{OB}$ have the same direction if the points A and B lying on the line $\overline{OAB}$ are on the same side of the point O. V ice versa, if A and B are on opposite sides of O, then $\boldsymbol{a}$ and $\boldsymbol{b}$ have opposite directions.

From the above, it follows that every vector is uniquely determined by its magnitude (modulus) and its direction.

Definition 3.3.2
n vectors are said to be ***coplanar*** if they lie in the same plane (that is, if they are parallel to the same plane) (Figure 3.12).

Of course, if $\{\boldsymbol{r}_1, \boldsymbol{r}_2, \ldots, \boldsymbol{r}_n\} \in \mathscr{V}(0)$, $n \geq 2$, is any (finite) set of radius-vectors such that $\boldsymbol{r}_i = \overrightarrow{OA_i}$, $i = 1, \ldots, n$, then $\boldsymbol{r}_1, \boldsymbol{r}_2, \ldots, \boldsymbol{r}_n$ are coplanar if the points $O, A_1, A_2, \ldots, A_n$ lie in the same plane (Figure 3.13).

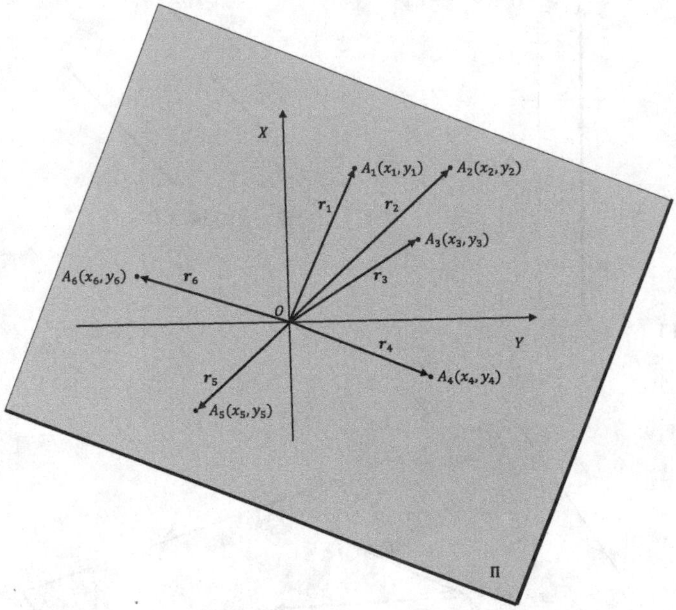

FIGURE 3.13

3.4 Addition of Vectors

Definition 3.4.1 (Addition of vectors)
Let $\mathscr{V}$ be a 3-dimensional space and $\boldsymbol{a}, \boldsymbol{b} \in \mathscr{V}$ be any two vectors. We call the binary operation

$$s : \mathscr{V} \times \mathscr{V} \to \mathscr{V}$$

defined as

$$s(\boldsymbol{a}, \boldsymbol{b}) = \boldsymbol{a} + \boldsymbol{b} = \boldsymbol{c}$$

the ***addition of vectors***. The sum $\boldsymbol{a}+\boldsymbol{b}=\boldsymbol{c}$ is formed by placing the tip of $\boldsymbol{a}$ on the tail of $\boldsymbol{b}$ and then joining the tail of $\boldsymbol{a}$ to the tip of $\boldsymbol{b}$. This is sometimes called the "tip-to-tail rule" (Figure 3.14).

To put it differently, given two vectors $\boldsymbol{a}$ and $\boldsymbol{b}$ we can always translate one of them, say $\boldsymbol{b}$, so that its initial point (tail) coincides with the terminal point (tip) of $\boldsymbol{a}$. Then the vector $\boldsymbol{c}$, with the same initial point as $\boldsymbol{a}$ and the same terminal point as $\boldsymbol{b}$, is considered as the sum of $\boldsymbol{a}$ and $\boldsymbol{b}$, i.e.

$$\boldsymbol{c} = \boldsymbol{a} + \boldsymbol{b}$$

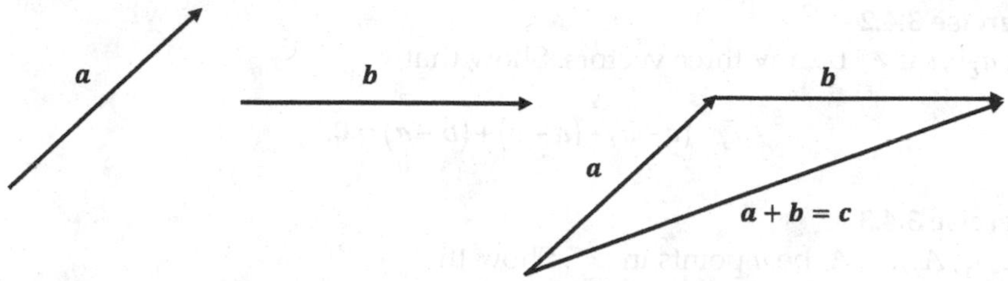

FIGURE 3.14

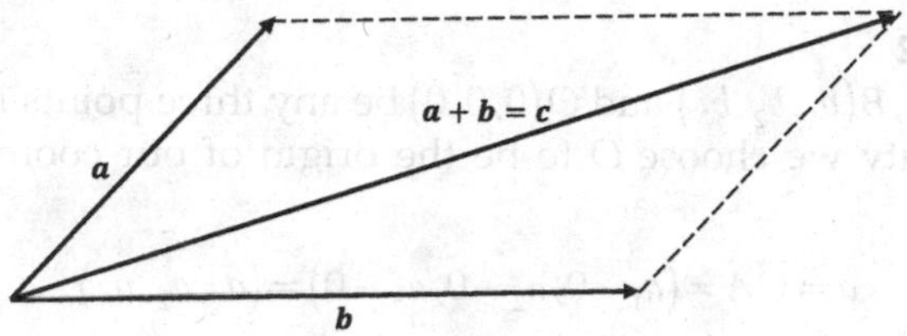

FIGURE 3.15(i)

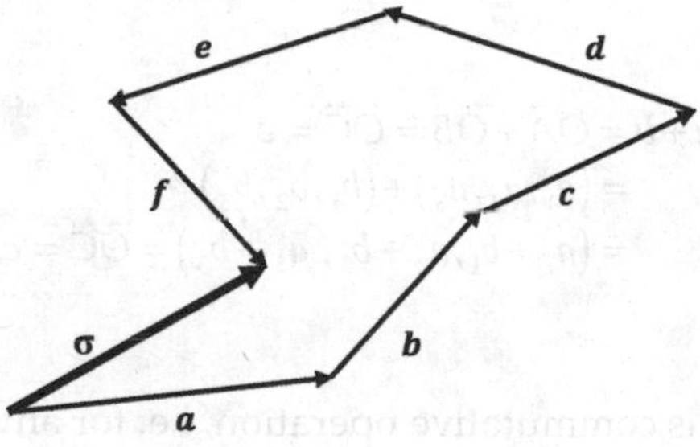

FIGURE 3.15(ii)

Of course, the same result would be obtained if, instead of placing the tip of $\boldsymbol{a}$ on the tail of $\boldsymbol{b}$, we place the tip of $\boldsymbol{b}$ on the tail of $\boldsymbol{a}$ and then join the tail of $\boldsymbol{b}$ to the tip of $\boldsymbol{a}$.

Yet another way to "visualize" the addition of vectors is to join the tails of $\boldsymbol{a}$ and $\boldsymbol{b}$ and form the corresponding parallelogram whose diagonal represents the sum $\boldsymbol{a}+\boldsymbol{b}$ (Figure 3.15(i)).

We can add n vectors following the same rule (Figure 3.15(ii)):

$$\boldsymbol{a}+\boldsymbol{b}+\boldsymbol{c}+\boldsymbol{d}+\boldsymbol{e}+\boldsymbol{f}=\boldsymbol{\sigma}.$$

Exercise 3.4.1

Let $\boldsymbol{a} \in \mathscr{V}$ be any vector. Show that

$$2a+a=3a.$$

Exercise 3.4.2
Let $\boldsymbol{a},\boldsymbol{b},\boldsymbol{c} \in \mathscr{V}$ be any three vectors. Show that

$$(\boldsymbol{c}-\boldsymbol{b})+(\boldsymbol{a}-\boldsymbol{c})+(\boldsymbol{b}-\boldsymbol{a})=\boldsymbol{0}.$$

Exercise 3.4.3
Let $A_1, A_2, \ldots, A_n$ be n points in $\mathscr{V}$. Show that

$$\overrightarrow{A_1A_2}+\overrightarrow{A_2A_3}+\ldots+\overrightarrow{A_{n-1}A_n}+\overrightarrow{A_nA_1}=\boldsymbol{0}.$$

Definition 3.4.2
Let $A(a_1,a_2,a_3), B(b_1,b_2,b_3)$ and $O(0,0,0)$ be any three points in $\mathscr{V}$ (without loss of generality we choose O to be the origin of our coordinate system). Then, with

$$\boldsymbol{a}=\overrightarrow{OA}=(a_1-0,a_2-0,a_3-0)=(a_1,a_2,a_3),$$

$$\boldsymbol{b}=\overrightarrow{OB}=(b_1-0,b_2-0,b_3-0)=(b_1,b_2,b_3),$$

we define

$$\begin{aligned}\boldsymbol{a}+\boldsymbol{b}&=\overrightarrow{OA}+\overrightarrow{OB}=\overrightarrow{OC}=\boldsymbol{c}\\&=(a_1,a_2,a_3)+(b_1,b_2,b_3)\\&=(a_1+b_1,a_2+b_2,a_3+b_3)=\overrightarrow{OC}=\boldsymbol{c}.\end{aligned}$$

Proposition 3.4.1
Addition of vectors is commutative operation, i.e. for any $\boldsymbol{a},\boldsymbol{b} \in \mathscr{V}$,

$$\boldsymbol{a}+\boldsymbol{b}=\boldsymbol{b}+\boldsymbol{a}.$$

Proof
Left to the reader. ■

Consequently, we see that, given any two vectors $\boldsymbol{a},\boldsymbol{b} \in \mathscr{V}$, by the difference $\boldsymbol{a}-\boldsymbol{b}$ we mean (Figure 3.16(i)):

$$\boldsymbol{a}-\boldsymbol{b}=\boldsymbol{a}+(-1)\boldsymbol{b}=\boldsymbol{a}+(-\boldsymbol{b}).$$

Another way to look at this is sketched in Figure 3.16(ii).

Definition 3.4.3
Let $\boldsymbol{a} \in \mathscr{V}$ be any non-zero vector. We say that $\boldsymbol{a}$ and $-\boldsymbol{a}$ are a pair of opposite vectors (additive inverses of each other), that is

$$\boldsymbol{a}+(-\boldsymbol{a})=(-\boldsymbol{a})+\boldsymbol{a}=\boldsymbol{0}.$$

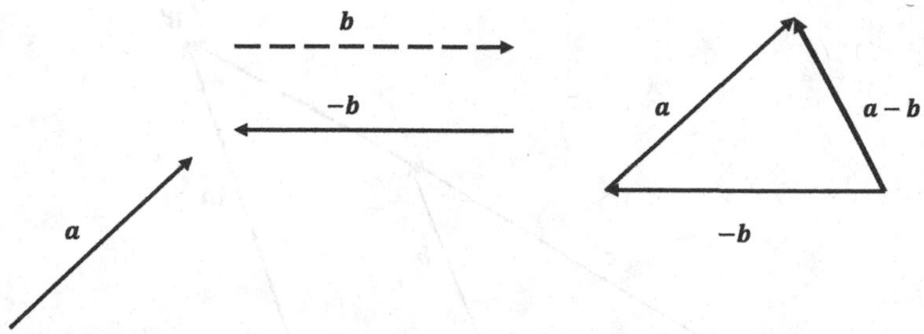

FIGURE 3.16(i)

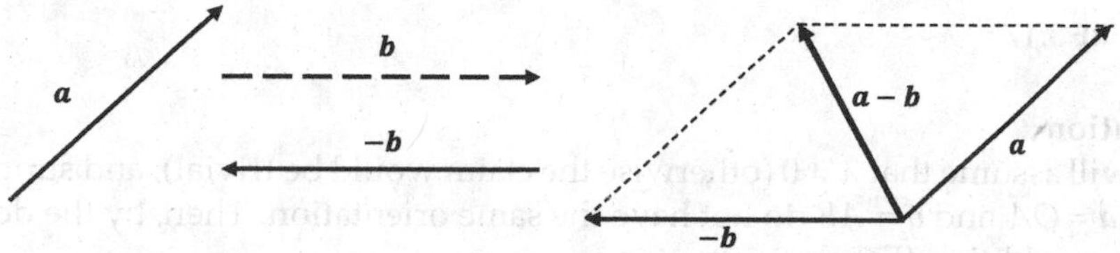

FIGURE 3.16(ii)

We say that the null-vector **0** is the *additive identity* (the neutral element with respect to addition).

Exercise 3.4.4
Show that

$$-(-\boldsymbol{a}) = \boldsymbol{a}.$$

Exercise 3.4.5
Let $\boldsymbol{a}, \boldsymbol{0} \in \mathscr{V}$ and $0 \in \mathbf{R}$. Show that

(i) $\boldsymbol{a} + \boldsymbol{0} = \boldsymbol{0} + \boldsymbol{a} = \boldsymbol{a}$;
(ii) $0 \cdot \boldsymbol{a} = \boldsymbol{0}$.

Exercise 3.4.6
Show that the sum of the vectors from the center of a regular pentagon to its vertices is a null-vector.

Example 3.4.1
Show that for any $\lambda \in \mathbf{R}$ and any $\boldsymbol{a}, \boldsymbol{b} \in \mathscr{V}$

$$\lambda(\boldsymbol{a} + \boldsymbol{b}) = \lambda \boldsymbol{a} + \lambda \boldsymbol{b}.$$

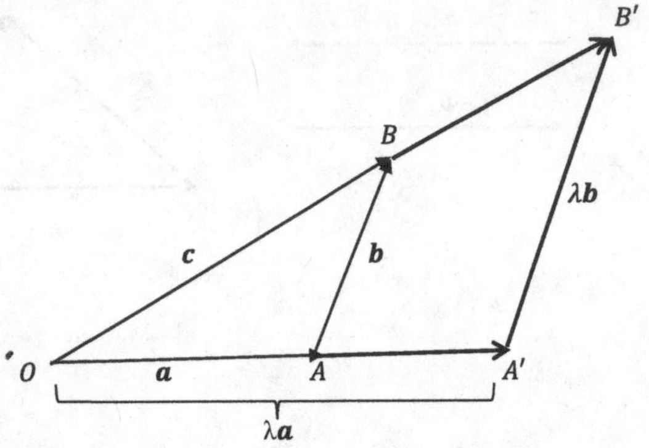

FIGURE 3.17

Solution
We will assume that $\lambda \neq 0$ (otherwise the claim would be trivial), and suppose that $a = \overrightarrow{OA}$ and $b = \overrightarrow{AB}$ do not have the same orientation. Then, by the definition of addition (Figure 3.17),

$$a+b=c=\overrightarrow{OB}.$$

Let $\lambda a = \overrightarrow{OA'}$. Notice that the points O, A and A' are collinear. If we choose the points B, B' and O to be collinear then the triangles ΔOAB and $\Delta OA'B'$ are similar. It follows that $\overline{AB} \,||\, \overline{A'B'}$, and therefore $\overrightarrow{A'B'} = \lambda\overrightarrow{AB} = \lambda b$.

On the other hand, the similarity of ΔOAB and $\Delta OA'B'$ also implies that $\overline{OB'} = \lambda\overline{OB}$, and thus $\overrightarrow{OB'} = \lambda\overrightarrow{OB} = \lambda b$. So we have

$$\lambda a + \lambda b = \overrightarrow{OA'} + \overrightarrow{A'B'} = \overrightarrow{OB'} = \lambda\overrightarrow{OB} = \lambda(a+b).$$ ■

Exercise 3.4.7
Show that

$$-(a+b) = -a-b.$$

Example 3.4.2
Show that

$$a-b = -(b-a).$$

Solution

$$\begin{aligned} a-b &= a+(-1)b \\ &= (-1)\cdot b + a \\ &= (-1)\cdot b + (-1)(-1)a \\ &= (-1)(b-a) \\ &= -(b-a). \end{aligned}$$ ■

Example 3.4.3
Show that for any $\lambda, \mu, \nu \in \mathbf{R}$ and any $\boldsymbol{a}, \boldsymbol{b}, \boldsymbol{c} \in \mathscr{V}$

$$(\lambda+\mu)(\boldsymbol{a}-\boldsymbol{b})-\lambda(\boldsymbol{a}+\boldsymbol{c})+(\mu-\nu)\boldsymbol{b}=\mu\boldsymbol{a}-(\lambda+\nu)\boldsymbol{b}-\lambda\boldsymbol{c}.$$

Solution

$$\begin{aligned}(\lambda+\mu)(\boldsymbol{a}-\boldsymbol{b})-\lambda(\boldsymbol{a}+\boldsymbol{c})+(\mu-\nu)\boldsymbol{b} &= (\lambda+\mu)\boldsymbol{a}-(\lambda+\mu)\boldsymbol{b}-\lambda\boldsymbol{a}-\lambda\boldsymbol{c}+\mu\boldsymbol{b}-\nu\boldsymbol{b}\\ &= \lambda\boldsymbol{a}+\mu\boldsymbol{a}-\lambda\boldsymbol{b}-\mu\boldsymbol{b}-\lambda\boldsymbol{a}-\lambda\boldsymbol{c}+\mu\boldsymbol{b}-\nu\boldsymbol{b}\\ &= \mu\boldsymbol{a}-\lambda\boldsymbol{b}-\nu\boldsymbol{b}-\lambda\boldsymbol{c}\\ &= \mu\boldsymbol{a}-(\lambda+\nu)\boldsymbol{b}-\lambda\boldsymbol{c}.\end{aligned}$$

■

Exercise 3.4.8
Let $\boldsymbol{a}=\overrightarrow{OA}$ and $\lambda \in \mathbf{R}$. Show that $|\lambda\boldsymbol{a}| = |\lambda||\boldsymbol{a}|$.

Exercise 3.4.9
Let $\boldsymbol{a}=\overrightarrow{OA}$ and $\boldsymbol{b}=\overrightarrow{OB}=\lambda\boldsymbol{a}$, with $\lambda \in \mathbf{R}$. Show that $\left|\overrightarrow{AB}\right| = |\boldsymbol{b}|-|\boldsymbol{a}|$.

As a simple exercise the reader should prove the following:

Proposition 3.4.2
If $\boldsymbol{a}, \boldsymbol{b} \in \mathscr{V}$ are non-zero vectors, then $\boldsymbol{a}$ is parallel to $\boldsymbol{b}$ iff there exists a non-zero scalar λ or μ such that $\boldsymbol{a}=\lambda\boldsymbol{b}$ or $\boldsymbol{b}=\mu\boldsymbol{a}$.

Example 3.4.4
Show that the line segment $\overline{MN}$ joining the midpoints of any two sides of a triangle ΔABC is parallel to the third side of the triangle (Figure 3.18).

Solution
Let $\overrightarrow{AB}=\boldsymbol{a}$, $\overrightarrow{AC}=\boldsymbol{b}$, $\overrightarrow{MN}=\boldsymbol{c}$ and $\overrightarrow{AN}=\boldsymbol{d}$, where M and N are the midpoints of the sides $\overline{AC}$ and $\overline{BC}$, respectively. Then,

$$\begin{aligned}\boldsymbol{d} &= \frac{1}{2}\boldsymbol{b}+\boldsymbol{c}\\ &= \boldsymbol{a}+\frac{1}{2}(\boldsymbol{b}-\boldsymbol{a})=\frac{1}{2}\boldsymbol{b}+\frac{1}{2}\boldsymbol{a}.\end{aligned}$$

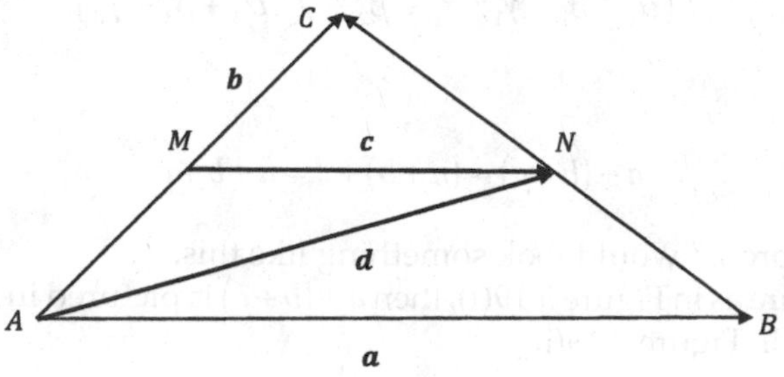

FIGURE 3.18

Hence $c = \frac{1}{2}a$, i.e., vectors a and c are parallel, i.e. line segments $\overline{AB}$ and $\overline{MN}$ are parallel. By the way, observe that the length of $\overline{MN}$ is equal to one-half of the length of $\overline{AB}$. ■

Proposition 3.4.3
Vector addition is associative, i.e.

$$a+(b+c)=(a+b)+c=a+b+c.$$

Proof
The proof is almost trivial. Let $O(0,0,0)$, $A(\alpha_1,\alpha_2,\alpha_3), B(\beta_1,\beta_2,\beta_3)$, $C(\gamma_1,\gamma_2,\gamma_3)$, with $\alpha_i,\beta_i,\gamma_i \in \mathbf{R}$, be any four points in $\mathscr{V}$. If we choose O to be the origin of the coordinate system, then

$$a=\overrightarrow{OA}=(a_1,a_2,a_3),\quad b=\overrightarrow{OB}=(\beta_1,\beta_2,\beta_3) \quad \text{and} \quad c=\overrightarrow{OC}=(\gamma_1,\gamma_2,\gamma_3).$$

Set

$$a+(b+c)=\overrightarrow{OT} \text{ and } (a+b)+c=\overrightarrow{OS}.$$

We would like to prove that $T = S$, i.e., that the coordinates of the points T and S are the same.

$$\begin{aligned} a+(b+c) &= (\alpha_1,\alpha_2,\alpha_3)+\left[(\beta_1,\beta_2,\beta_3)+(\gamma_1,\gamma_2,\gamma_3)\right] \\ &= (\alpha_1,\alpha_2,\alpha_3)+(\beta_1+\gamma_1,\ \beta_2+\gamma_2,\ \beta_3+\gamma_3) \\ &= (\alpha_1+\beta_1+\gamma_1,\ \alpha_2+\beta_2+\gamma_2,\alpha_3+\beta_3+\gamma_3)=\overrightarrow{OT}. \end{aligned}$$

$$\begin{aligned} (a+b)+c &= \left[(\alpha_1+\beta_1,\alpha_2+\beta_2,\alpha_3+\beta_3)\right]+(\gamma_1,\gamma_2,\gamma_3) \\ &= (\alpha_1+\beta_1+\gamma_1,\ \alpha_2+\beta_2+\gamma_2,\ \alpha_3+\beta_3+\gamma_3)=\overrightarrow{OS}. \end{aligned}$$

So, the coordinates of T and S are the same, i.e.

$$\begin{aligned} &S(\alpha_1+\beta_1+\gamma_1,\alpha_2+\beta_2+\gamma_2,\ \alpha_3+\beta_3+\gamma_3)= \\ &T(\alpha_1+\beta_1+\gamma_1,\ \alpha_2+\beta_2+\gamma_2,\alpha_3+\beta_3+\gamma_3). \end{aligned}$$

Hence

$$a+(b+c)=(a+b)+c=a+b+c.$$

A graphical "proof" would look something like this:

If a, b and c are as in Figure 3.19(i), then $a+(b+c)$ is pictured in Figure 3.19(ii) and $(a+b)+c$ in Figure 3.19(iii).

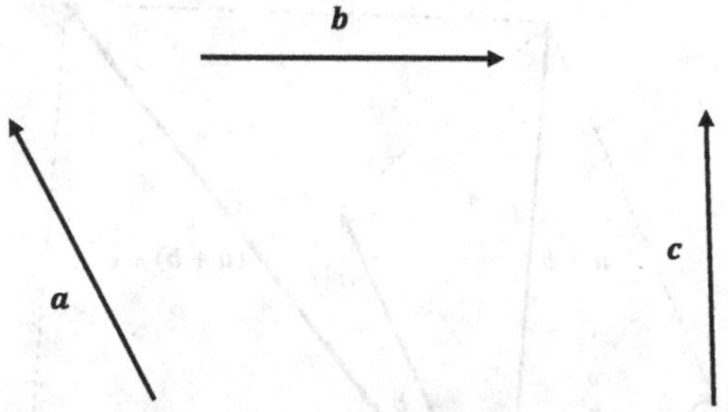

FIGURE 3.19(i)

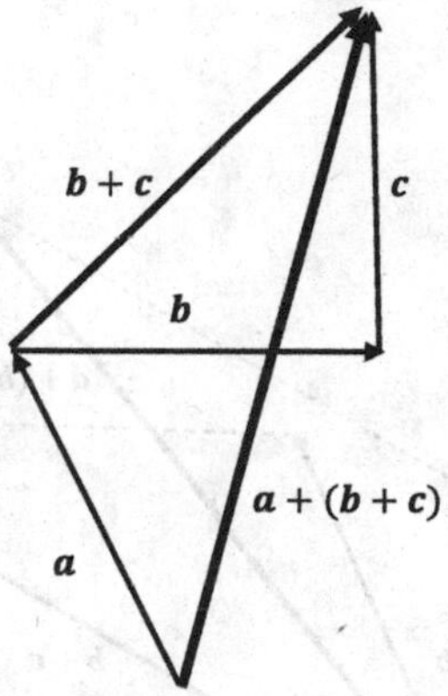

FIGURE 3.19(ii)

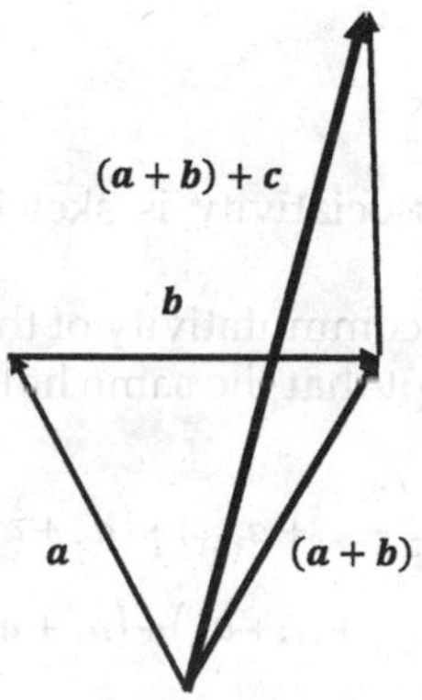

FIGURE 3.19(iii)

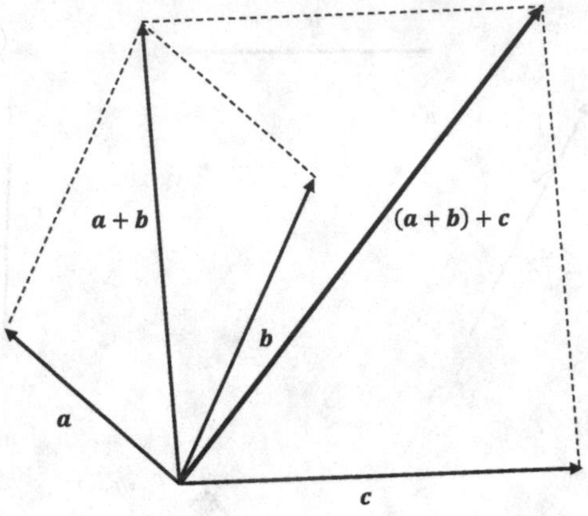

FIGURE 3.20(i)

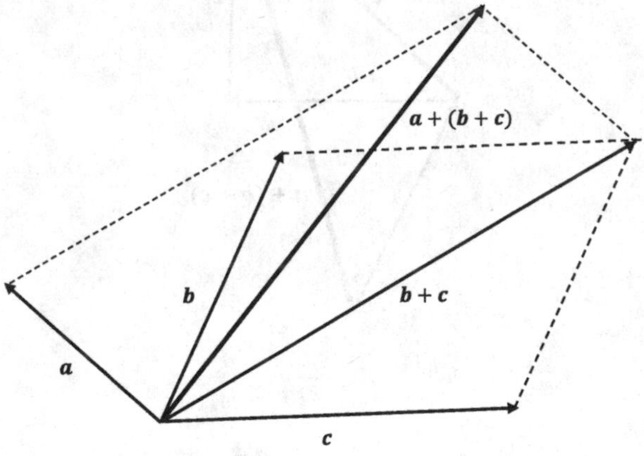

FIGURE 3.20(ii)

Another way to depict associativity is sketched in the following figure (Figure 3.4.20(i), (ii)). ■

Once the associativity and commutativity of the addition of three vectors is established, it is easy to accept that the same holds for n vectors, i.e.

$$\sum_{i=1}^{n} a_i = (a_1 + a_2 + \ldots + a_{k-1}) + (a_k + a_{k+1} + \ldots + a_n)$$
$$= (a_k + a_{k+1} + \ldots + a_n) + (a_1 + a_2 + \ldots + a_{k-1}).$$

Example 3.4.5
Show that

$$(a+b)+(c+d) = [(b+d)+a]+c.$$

Solution

$$\begin{aligned}(a+b)+(c+d) &= (a+b)+(d+c)\\ &= \left[(a+b)+d\right]+c\\ &= \left[a+(b+d)\right]+c\\ &= \left[(b+d)+a\right]+c.\end{aligned}$$

■

Example 3.4.6
Show that

$$a-(b+c)=(a-b)-c.$$

Solution

$$\begin{aligned}a-(b+c) &= a+(-1)(b+c)\\ &= a+\left[(-1)b+(-1)c\right]\\ &= \left[a+(-1)b\right]+(-1)c\\ &= (a-b)-c.\end{aligned}$$

■

Example 3.4.7
Let M be the midpoint of a segment $\overline{AB}$, and let O be any point in space (Figure 3.21). Show that

$$\overrightarrow{OM}=\frac{1}{2}\overrightarrow{OA}+\frac{1}{2}\overrightarrow{OB}.$$

Solution
From Figure 3.21 we see that

$$\overrightarrow{OA}+\overrightarrow{AB}=\overrightarrow{OB}$$

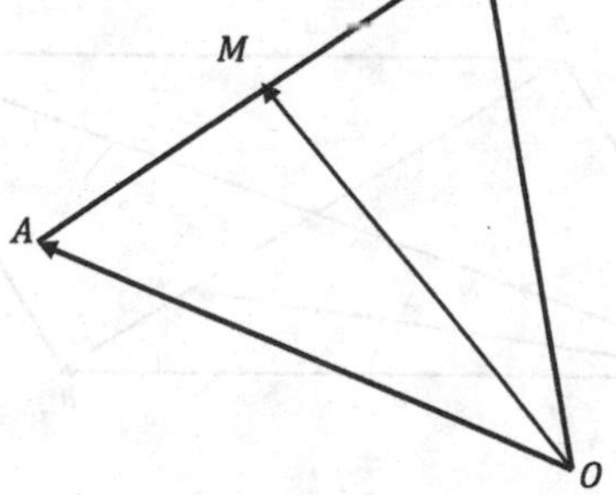

FIGURE 3.21

$$\overrightarrow{OM}+\frac{1}{2}\overrightarrow{AB}=\overrightarrow{OB}$$

$$\overrightarrow{OA}+\frac{1}{2}\overrightarrow{AB}=\overrightarrow{OM}.$$

Thus

$$\overrightarrow{OM}=\overrightarrow{OB}-\frac{1}{2}\overrightarrow{AB}$$

and

$$\overrightarrow{OM}=\overrightarrow{OA}+\frac{1}{2}\overrightarrow{AB}$$

So

$$2\overrightarrow{OM}=\overrightarrow{OA}+\overrightarrow{OB}$$

Hence

$$\overrightarrow{OM}=\frac{1}{2}\overrightarrow{OA}+\frac{1}{2}\overrightarrow{OB}.$$ ■

Example 3.4.8
Show that the diagonals of a parallelogram bisect each other.

Solution
Consider the parallelogram *ABCD* shown in Figure 3.22,

where $\boldsymbol{a}=\overrightarrow{AB}$ and $\boldsymbol{b}=\overrightarrow{AD}$. If the diagonals of a parallelogram bisect each other, and *M* and *N* are the midpoints of $\overline{AC}$ and $\overline{BD}$, respectively, then *M* and *N* have to coincide, i.e. $M=N$. If *M* and *N* are the midpoints of *AC* and *BD*, then

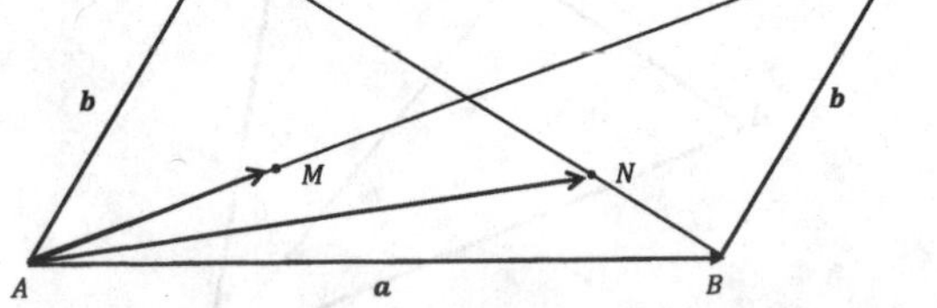

FIGURE 3.22

$$\overrightarrow{AM} = \frac{1}{2}\overrightarrow{AC},$$

and

$$\overrightarrow{AN} = \frac{1}{2}\boldsymbol{a} + \frac{1}{2}\boldsymbol{b}.$$

But

$$\overrightarrow{AC} = \boldsymbol{a} + \boldsymbol{b}.$$

Thus,

$$\overrightarrow{AM} = \frac{1}{2}\overrightarrow{AC} = \frac{1}{2}(\boldsymbol{a} + \boldsymbol{b}) = \overrightarrow{AN}.$$

Therefore, $M = N$. ■

Proposition 3.4.4

The points A, B and P are collinear iff, for any point O in space, $\overrightarrow{OP} = (1-m)\overrightarrow{OA} + m\overrightarrow{OB}$.

Proof

Let O, A, B and P be points as shown in Figure 3.23, such that P divides the line segment $\overline{AB}$ in the ratio $m : n$. Then,

$$\begin{aligned}\overrightarrow{OP} &= \overrightarrow{OA} + \overrightarrow{AP} = \overrightarrow{OA} + m\overrightarrow{AB}\\ &= \overrightarrow{OA} + m\left(\overrightarrow{OB} - \overrightarrow{OA}\right)\\ &= (1-m)\overrightarrow{OA} + m\overrightarrow{OB}.\end{aligned}$$

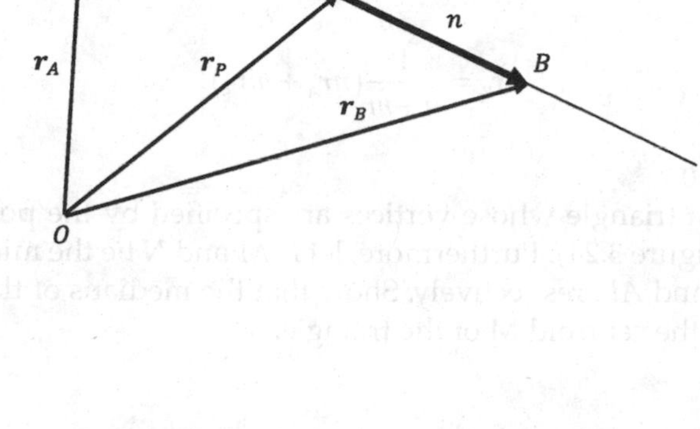

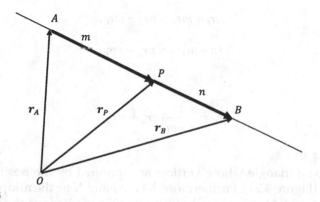

FIGURE 3.23

Conversely, if

$$\overrightarrow{OP} = (1-m)\overrightarrow{OA} + m\overrightarrow{OB},$$

Then

$$\overrightarrow{OP} - \overrightarrow{OA} = m\left(\overrightarrow{OB} - \overrightarrow{OA}\right),$$

i.e.

$$\overrightarrow{AP} = m\overrightarrow{AB}.$$

Thus, $\overrightarrow{AP}$ is collinear to $\overrightarrow{AB}$, so the line segment $\overline{AP}$ lies along line segment $\overline{AB}$, and therefore points A, B and P are collinear. ■

Example 3.4.9
Let P divide AB in the ratio $m:n$, with $m,n>0$, and let O be any point in space (Figure 3.23) with $\overrightarrow{OA} = \boldsymbol{r}_A$, $\overrightarrow{OP} = \boldsymbol{r}_P$ and $\overrightarrow{OB} = \boldsymbol{r}_B$. Show that

$$\boldsymbol{r}_P = \frac{1}{m+n}\left(n\boldsymbol{r}_A + m\boldsymbol{r}_B\right).$$

Solution
Let $\overrightarrow{OA} = \boldsymbol{r}_A$, $\overrightarrow{OP} = \boldsymbol{r}_P$ and $\overrightarrow{OB} = \boldsymbol{r}_B$.
Then
$\overrightarrow{AP} = \boldsymbol{r}_P - \boldsymbol{r}_A$ and $\overrightarrow{PB} = \boldsymbol{r}_B - \boldsymbol{r}_P$.
On the other hand, by our assumption, $\overrightarrow{AP} = \left(m/n\right)\overrightarrow{PB}$, so

$$n\left(\boldsymbol{r}_P - \boldsymbol{r}_A\right) = m\left(\boldsymbol{r}_B - \boldsymbol{r}_P\right).$$

It follows that

$$n\boldsymbol{r}_P + m\boldsymbol{r}_P = n\boldsymbol{r}_A + m\boldsymbol{r}_B,$$

$$(n+m)\boldsymbol{r}_P = n\boldsymbol{r}_A + m\boldsymbol{r}_B.$$

Thus

$$\boldsymbol{r}_P = \frac{1}{n+m}\left(n\boldsymbol{r}_A + m\boldsymbol{r}_B\right).$$ ■

Example 3.4.10
Let ΔABC be a triangle whose vertices are specified by the position vectors $\boldsymbol{r}_A, \boldsymbol{r}_B$ and $\boldsymbol{r}_C$ (Figure 3.24). Furthermore, let L, M and N be the midpoints of the sides BC, CA and AB, respectively. Show that the medians of the triangle are concurrent at the centroid M of the triangle.

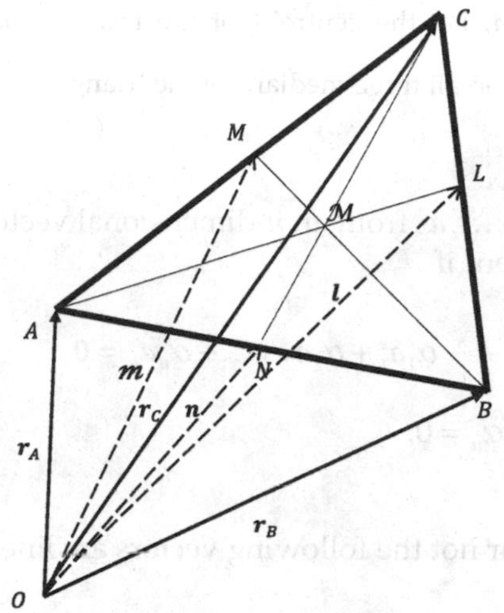

FIGURE 3.24

Solution

From the figure we see that L has the position vector $\boldsymbol{l} = \frac{1}{2}(\boldsymbol{r}_B + \boldsymbol{r}_C)$, M has the position vector $\boldsymbol{m} = \frac{1}{2}(\boldsymbol{r}_A + \boldsymbol{r}_C)$ and N has the position vector $\boldsymbol{n} = \frac{1}{2}(\boldsymbol{r}_A + \boldsymbol{r}_B)$. Now the point which divides $\overline{AL}$ in the ratio 2 : 1, according to the result from previous example, has the position vector

$$\begin{aligned}\frac{1}{2+1}(1\cdot \boldsymbol{r}_A + 2\cdot \boldsymbol{l}) &= \frac{1}{3}\left[\boldsymbol{r}_A + 2\cdot\frac{1}{2}(\boldsymbol{r}_B + \boldsymbol{r}_C)\right]\\ &= \frac{1}{3}(\boldsymbol{r}_A + \boldsymbol{r}_B + \boldsymbol{r}_C).\end{aligned}$$

The point which divides $\overline{BM}$ in the ratio 2 : 1 has the position vector

$$\begin{aligned}\frac{1}{2+1}(1\cdot \boldsymbol{r}_B + 2\cdot \boldsymbol{m}) &= \frac{1}{3}\left[\boldsymbol{r}_B + 2\cdot\frac{1}{2}(\boldsymbol{r}_A + \boldsymbol{r}_C)\right]\\ &= \frac{1}{3}(\boldsymbol{r}_A + \boldsymbol{r}_B + \boldsymbol{r}_C).\end{aligned}$$

Thus, the point which divides $\overline{CN}$ in the ratio 2 : 1 has the position vector

$$\begin{aligned}\frac{1}{2+1}(1\cdot \boldsymbol{r}_C + 2\cdot \boldsymbol{n}) &= \frac{1}{3}\left[\boldsymbol{r}_C + 2\cdot\frac{1}{2}(\boldsymbol{r}_A + \boldsymbol{r}_B)\right]\\ &= \frac{1}{3}(\boldsymbol{r}_A + \boldsymbol{r}_B + \boldsymbol{r}_C).\end{aligned}$$

Therefore, the point $\mathcal{M}$, i.e. the centroid of the triangle, has the position vector $\frac{1}{3}(\boldsymbol{r}_A + \boldsymbol{r}_B + \boldsymbol{r}_C)$ and lies on all three medians of the triangle. ■

Definition 3.4.4
A set of vectors $\boldsymbol{a}_1, \boldsymbol{a}_2, \ldots, \boldsymbol{a}_n$ from an n-dimensional vector space $\mathscr{V}$ is said to be linearly independent if

$$\alpha_1 \boldsymbol{a}_1 + \alpha_2 \boldsymbol{a}_2 + \ldots + \alpha_n \boldsymbol{a}_n = \boldsymbol{0}$$

only if $\alpha_1 = \alpha_2 = \ldots = \alpha_n = 0$.

Exercise 3.4.10
Determine whether or not the following vectors are linearly independent:

(i) $\boldsymbol{a} = (\pi, 0), \boldsymbol{b} = (0, 1)$;
(ii) $\boldsymbol{a} = (0,1,1), \boldsymbol{b} = (0,2,1), \boldsymbol{c} = (1,5,3)$.

Exercise 3.4.11
Determine whether the vectors

$$\boldsymbol{x} = \boldsymbol{a} + 2\boldsymbol{b} + \boldsymbol{c}, \boldsymbol{y} = \boldsymbol{a} + 3\boldsymbol{b} - 2\boldsymbol{c} \text{ and } \boldsymbol{z} = \boldsymbol{a} + \boldsymbol{b} + 4\boldsymbol{c}$$

are linearly dependent or independent.

Exercise 3.4.12
Let $\boldsymbol{a}$ and $\boldsymbol{b}$ be two linearly independent vectors, and let

$$\boldsymbol{c} = \alpha \boldsymbol{a} + \beta \boldsymbol{b}$$

where $\alpha, \beta \in \mathbf{R}$. Express α and β in terms of $\boldsymbol{a}, \boldsymbol{b}$ and $\boldsymbol{c}$.

Definition 3.4.5
Let $\boldsymbol{c} \in \mathscr{V}$ be any non-zero vector such that

$$\boldsymbol{c} = \alpha_1 \boldsymbol{a}_1 + \alpha_2 \boldsymbol{a}_2 + \ldots + \alpha_n \boldsymbol{a}_n, \ \boldsymbol{a}_i \in \mathscr{V}, \alpha_i \in \mathbf{R}.$$

We say that $\boldsymbol{c}$ is a linear combination of vectors $\boldsymbol{a}_1, \boldsymbol{a}_2, \ldots, \boldsymbol{a}_n$.

Proposition 3.4.5
Let $\boldsymbol{a}, \boldsymbol{b} \in \mathscr{V}$ be any two vectors, and let $\alpha, \beta \in \mathbf{R}$, such that

$$\boldsymbol{c} = \alpha \boldsymbol{a} + \beta \boldsymbol{b}.$$

Then $\boldsymbol{a}, \boldsymbol{b}$ and $\boldsymbol{c}$ are coplanar (Figure 3.25).

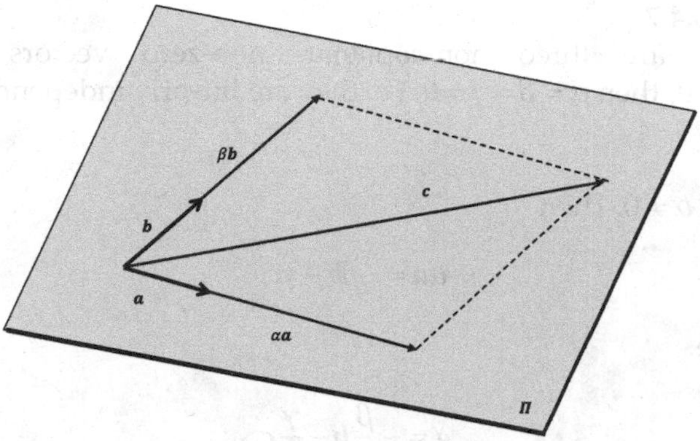

FIGURE 3.25

Proof
The proof follows immediately from the definition of vector addition. ■

Proposition 3.4.5'
Let $\boldsymbol{a},\boldsymbol{b}\in\mathscr{V}$ be any two non-collinear vectors, and let $\boldsymbol{c}\in\mathscr{V}$ be a vector that is coplanar with $\boldsymbol{a}$ and $\boldsymbol{b}$. Then there exist $\alpha,\beta\in\mathbf{R}$ such that

$$\boldsymbol{c}=\alpha\boldsymbol{a}+\beta\boldsymbol{b}.$$

Subsequently, we have

Proposition 3.4.6
If $\boldsymbol{a},\boldsymbol{b}\in\mathscr{V}$ are two non-zero, non-collinear vectors, such that $\alpha\boldsymbol{a}+\beta\boldsymbol{b}=0$, with $\alpha,\beta\in\mathbf{R}$, then $\alpha=\beta=0$.

Proof
Suppose $\alpha\neq 0$. Then

$$\alpha\boldsymbol{a}=-\beta\boldsymbol{b},$$

and therefore

$$\boldsymbol{a}=-\frac{\beta}{\alpha}\boldsymbol{b}.$$

implying that $\boldsymbol{a}$ and $\boldsymbol{b}$ are collinear, contrary to our hypothesis. Thus $\alpha=0$, and therefore

$$\beta\boldsymbol{b}=0.$$

Since $\boldsymbol{b}\neq\mathbf{0}$, it follows that $\beta=0$. ■

Proposition 3.4.7
If $\boldsymbol{a},\boldsymbol{b},\boldsymbol{c}\in\mathscr{V}$ are three non-coplanar, non-zero vectors such that $\alpha\boldsymbol{a}+\beta\boldsymbol{b}+\gamma\boldsymbol{c}=\boldsymbol{0}$, then $\alpha=\beta=\gamma=0$, i.e. they are linearly independent.

Proof
Suppose that $\alpha\neq 0$. Then

$$\alpha\boldsymbol{a}=-\beta\boldsymbol{b}-\gamma\boldsymbol{c},$$

and therefore

$$\boldsymbol{a}=-\frac{\beta}{\alpha}\boldsymbol{b}-\frac{\gamma}{\alpha}\boldsymbol{c}.$$

implying that $\boldsymbol{a}$ is coplanar with $\boldsymbol{b}$ and $\boldsymbol{c}$. But that contradicts our original hypothesis that $\boldsymbol{a},\boldsymbol{b}$ and $\boldsymbol{c}$ are non-coplanar. Hence $\alpha=0$, and $\beta\boldsymbol{b}+\gamma\boldsymbol{c}=\boldsymbol{0}$. By Proposition 3.4.6, $\beta=\gamma=0$. So, indeed, $\alpha=\beta=\gamma=0$.

Note also that, given three *non-coplanar* vectors $\boldsymbol{a},\boldsymbol{b},\boldsymbol{c}\in\mathscr{V}$, no two of them can be collinear. ■

Proposition 3.4.8
If $\boldsymbol{a},\boldsymbol{b},\boldsymbol{c}\in\mathscr{V}$ are three non-zero and non-coplanar vectors, and if $\boldsymbol{x}$ is any other vector in $\mathscr{V}$, then $\boldsymbol{x}$ can be expressed as a linear combination of $\boldsymbol{a}$, $\boldsymbol{b}$, and $\boldsymbol{c}$, i.e.

$$\boldsymbol{x}=\alpha\boldsymbol{a}+\beta\boldsymbol{b}+\gamma\boldsymbol{c},\alpha,\beta,\gamma\in\mathbf{R}.$$ ■

Proof
If $\boldsymbol{x}=\boldsymbol{0}$ there is nothing to prove – by the previous proposition $\alpha=\beta=\gamma=0$ and the claim of the proposition is obvious. If $\boldsymbol{x}\neq\boldsymbol{0}$ and $\alpha=0$, then $\boldsymbol{x}=\beta\boldsymbol{b}+\gamma\boldsymbol{c}$. This means that $\boldsymbol{x}$ is coplanar with $\boldsymbol{b}$ and $\boldsymbol{c}$, contrary to our original hypothesis. Thus, $\alpha\neq 0$. Similarly we show that $\beta\neq 0$ and $\gamma\neq 0$. Hence,

$$\boldsymbol{x}=\alpha\boldsymbol{a}+\beta\boldsymbol{b}+\gamma\boldsymbol{c}.$$

3.5 Basis of a Vector Space

Definition 3.5.1
Let $\boldsymbol{B}=\{\boldsymbol{b}_1,\boldsymbol{b}_2,\boldsymbol{b}_3\}$ be an ordered triple of non-coplanar vectors from some 3-dimensional space $\mathscr{V}$. We say that $\boldsymbol{B}$ is a ***basis*** of the space $\mathscr{V}$ if any $\boldsymbol{a}\in\mathscr{V}$ can be *uniquely* represented as a linear combination of vectors from $\boldsymbol{B}$, i.e.

$$\boldsymbol{a}=\alpha_1\boldsymbol{b}_1+\alpha_2\boldsymbol{b}_2+\alpha_3\boldsymbol{b}_3,\quad \alpha_i\in\mathbf{R}.$$

Since the set of any three non-coplanar vectors from $\mathscr{V}$ forms a basis for $\mathscr{V}$ it is evident that there are infinitely many bases of $\mathscr{V}$. However, once a basis $\boldsymbol{B}$ is selected, the *components* $\alpha_1, \alpha_2, \alpha_3$ of the vector $\boldsymbol{a}$ are unique with respect to this basis, and we often represent the vector by simply writing $\boldsymbol{a} = (\alpha_1, \alpha_2, \alpha_3)$.

Proposition 3.5.1
if $\boldsymbol{a}, \boldsymbol{b}, \boldsymbol{c}$ are three *linearly independent* vectors from a 3-dimensional space $\mathscr{V}$, then any vector $\boldsymbol{x} \in \mathscr{V}$ can be *uniquely* expressed as a linear combination of $\boldsymbol{a}, \boldsymbol{b}$ and $\boldsymbol{c}$. This makes the set $\{\boldsymbol{a}, \boldsymbol{b}, \boldsymbol{c}\}$ a basis for $\mathscr{V}$.

Proof
By Proposition 3.4.8, any vector $\boldsymbol{x} \in \mathscr{V}$ can be expressed as a linear combination of $\boldsymbol{a}$,$\boldsymbol{b}$, and $\boldsymbol{c}$. Suppose that there are two representations of the vector $\mathbf{x}$,

$$\boldsymbol{x} = \alpha_1\boldsymbol{a} + \beta_1\boldsymbol{b} + \gamma_1\boldsymbol{c} = \alpha_2\boldsymbol{a} + \beta_2\boldsymbol{b} + \gamma_2\boldsymbol{c}.$$

Then,

$$(\alpha_1 - \alpha_2)\boldsymbol{a} + (\beta_1 - \beta_2)\boldsymbol{b} + (\gamma_1 - \gamma_2)\boldsymbol{c} = \mathbf{0}.$$

Since $\boldsymbol{a}$,$\boldsymbol{b}$, and $\boldsymbol{c}$ are linearly independent,

$$\alpha_1 - \alpha_2 = 0,$$
$$\beta_1 - \beta_2 = 0,$$
$$\gamma_1 - \gamma_2 = 0,$$

i.e.

$$\alpha_1 = \alpha_2,$$
$$\beta_1 = \beta_2,$$
$$\gamma_1 = \gamma_2.$$

Hence, $\boldsymbol{x}$ is uniquely represented as a linear combination of vectors $\boldsymbol{a}, \boldsymbol{b}, \boldsymbol{c}$, and the set $\{\boldsymbol{a}, \boldsymbol{b}, \boldsymbol{c}\}$ is a basis for $\mathscr{V}$. ■

Corollary 3.5.1
Any set $\{\boldsymbol{a}, \boldsymbol{b}, \boldsymbol{c}\}$ of three non-coplanar non-zero vectors in 3-dimensional $\mathscr{V}$ is a basis of $\mathscr{V}$.

Corollary 3.5.2
(i) Let $\mathscr{S} \subseteq \mathscr{V}$ be a set containing the zero vector. Then $\mathscr{S}$ is linearly dependent.
(ii) Two vectors $\boldsymbol{a}, \boldsymbol{b} \in \mathscr{V}$ are linearly independent iff they are non-collinear.

(iii) Three vectors $a, b, c \in \mathscr{V}$ are linearly independent iff they are non-coplanar.

(iv) If $\mathscr{A} \subseteq \mathscr{B}$ is a linearly dependent set of vectors, then $\mathscr{B}$ is also a linearly dependent set.

(v) If $\mathscr{B}$ is a linearly independent set of vectors and $\mathscr{A} \subseteq \mathscr{B}$, then $\mathscr{A}$ is also a linearly independent set.

Example 3.5.1

Suppose we choose the set $E = \{e_1 = (1,0,0), e_2 = (0,1,0), e_3 = (0,0,1)\}$. Readers can easily convince themselves that E is a basis of the 3-dimensional space $\mathscr{V}$, and this is called the ***canonical basis***. Then, some vector, say, $x = \left(2, \frac{2}{3}, -5\right)$ expressed in the basis E is

$$x = x_1 e_1 + x_2 e_2 - x_3 e_3 = 2e_1 + \frac{2}{3} e_2 - 5e_3. \quad \blacksquare$$

Proposition 3.5.2

Let $B = \{b_1, b_3, b_3\}$ be a given basis for $\mathscr{V}$, and let $x, y \in \mathscr{V}$ be any two vectors whose representations in basis B are

$$\begin{aligned} x &= x_1 b_1 + x_2 b_2 + x_3 b_3 \\ &= (x_1, x_2, x_3), x_i \in \mathbf{R}, \\ y &= y_1 b_1 + y_2 b_2 + y_3 b_3 \\ &= (y_1, y_2, y_3), y_i \in \mathbf{R}. \end{aligned}$$

Then

$$\begin{aligned} \text{(i)} \quad x + y &= (x_1, x_2, x_3) + (y_1, y_2, y_3) \\ &= (x_1 + y_1) b_1 + (x_2 + y_2) b_2 + (x_3 + y_3) b_3 \\ &= (x_1 + y_1, x_2 + y_2, x_3 + y_3). \end{aligned}$$

$$\begin{aligned} \text{(ii)} \quad \lambda x &= \lambda(x_1, x_2, x_3) \\ &= \lambda(x_1 b_1 + x_2 b_2 + x_3 b_3) = \lambda x_1 b_1 + \lambda x_2 b_2 + \lambda x_3 b_3 \\ &= (\lambda x_1, \lambda x_2, \lambda x_3), \forall \lambda \in \mathbf{R}. \end{aligned}$$

In particular,

$$-x = (-1)x = (-x_1, -x_2, -x_3).$$

Notice, also, that two vectors $x = (x_1, x_2, x_3)$ and $y = (y_1, y_2, y_3)$ are collinear if

$$x_1 : x_2 : x_3 = y_1 : y_2 : y_3.$$

In general, if $\boldsymbol{B} = \{\boldsymbol{b}_1, \boldsymbol{b}_2, \dots, \boldsymbol{b}_n\}$ is a basis of an n-dimensional vector space $\boldsymbol{X}$, and $\boldsymbol{x}_i \in \boldsymbol{X}, i = 1, \dots m$ are any vectors such that

$$\boldsymbol{x}_i = \alpha_{i1}\boldsymbol{b}_1 + \alpha_{i2} + \dots + \alpha_{in}\boldsymbol{b}_n = \sum_{j=1}^{n} \alpha_{ij}\boldsymbol{b}_j, \alpha_{ij} \in \Phi,$$

then

$$\sum_{i=1}^{m} \boldsymbol{x}_i = \boldsymbol{x}_1 + \boldsymbol{x}_2 + \dots + \boldsymbol{x}_m = \sum_{i=1}^{m}\sum_{j=1}^{n} \alpha_{ij}\boldsymbol{b}_j.$$

Similarly, if

$$\boldsymbol{x} = \sum_{i=1}^{n} a_i \boldsymbol{b}_i, \alpha_i \in \Phi,$$

then

$$\lambda \boldsymbol{x} = \lambda \sum_{i=1}^{n} \alpha_i \boldsymbol{b}_i = \sum_{i=1}^{n} \lambda \alpha_i \boldsymbol{b}_i.$$

Exercise 3.5.1

(i) Express $\boldsymbol{x} = (4, 3)$ as a linear combination of the vectors $\boldsymbol{b}_1 = (2, 1)$ and $\boldsymbol{b}_2 = (-1, 0)$.

(ii) Express $\boldsymbol{x} = (1, 1, 1)$ as a linear combination of the vectors $\boldsymbol{b}_1 = (0, 1, -1)$, $\boldsymbol{b}_2 = (1, 1, 0)$ and $\boldsymbol{b}_3 = (1, 0, 2)$.

Exercise 3.5.2

Let $\boldsymbol{b}_1$ and $\boldsymbol{b}_2$ be basis vectors of some 2-dimensional vector space. Find another set of basis vectors, say, $\boldsymbol{b}_1'$ and $\boldsymbol{b}_2'$.

Note

1 By the "magnitude" (or "modulus") of a vector $\boldsymbol{a} = \overrightarrow{AB}$, we mean the "length" of $\boldsymbol{a}$, i.e. the number obtained by measuring the length of $\overrightarrow{AB}$, and we write $|\overrightarrow{AB}| = |\boldsymbol{a}| = a$. The precise definition of "magnitude" will be given shortly.

4

Vectors in R^3 Space

Before one embarks on the study of the highly axiomatized n-dimensional (vector) algebra, the study of vectors in the 3-dimensional space $\mathbf{R}^3$ may prove to be useful, as the 3-dimensional model will serve nicely to give a "feel" for the higher-dimensional analogues of $\mathbf{R}^3$.

Indeed, if we replace the abstract space $\mathscr{V}$ from the previous chapter by $\mathbf{R}^3$, all the axioms of a vector space are satisfied, and we are justified in incorporating all the operations from $\mathscr{V}$ into $\mathbf{R}^3$.

4.1 $\{i, j, k\}$-basis of $\mathbf{R}^3$ Space

For convenience we will use the *right-handed rectangular Cartesian coordinate system*. In this system we choose as our basis a particular set of vectors $\boldsymbol{i}, \boldsymbol{j}$, and $\boldsymbol{k}$, having directions of the positive X, Y, and Z axes with respective components $(1,0,0),(0,1,0),(0,0,1)$, i.e.

$$\begin{aligned} \boldsymbol{i} &= (1,0,0), \\ \boldsymbol{j} &= (0,1,0), \\ \boldsymbol{k} &= (0,0,1). \end{aligned}$$

We call the set $\mathbf{B} = \{\mathbf{i}, \mathbf{j}, \mathbf{k}\}$ the ***orthonormal basis*** (***canonical basis***) of the 3-dimensional space R^3.

By right-handedness we mean that a rotation by an angle from $\boldsymbol{i}$ to $\boldsymbol{j}$ (indicated by the rotation arrow in Figure 4.1), accompanied by a translation in the direction of $\boldsymbol{k}$, gives a right-hand screw motion.

DOI: 10.1201/9781003343486-4

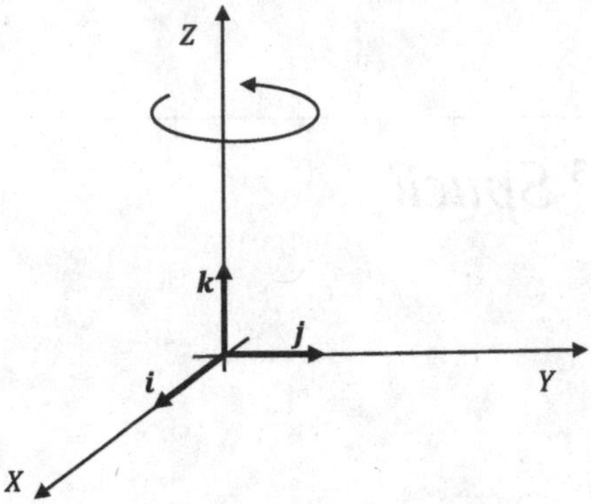

FIGURE 4.1

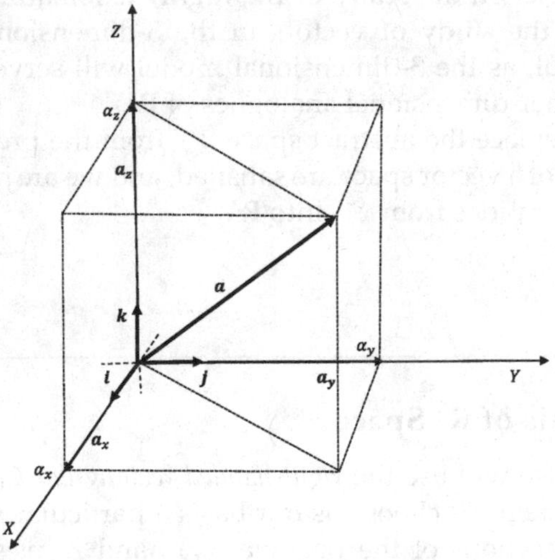

FIGURE 4.2

Proposition 4.1.1
The unit vectors $\boldsymbol{i}, \boldsymbol{j}$ and $\boldsymbol{k}$ are linearly independent and they span the space $\mathbf{R}^3$.

Exercise 4.1.1
Prove Proposition 4.1.1.

Definition 4.1.1
If $\boldsymbol{a}$ is a vector with its initial point – its tail – at the origin of the Cartesian coordinate system, and $(\alpha_x, \alpha_y, \alpha_z)$ are the coordinates of its terminal point (its tip) (Figure 4.2), then $\boldsymbol{a}$ is *uniquely* represented as a linear combination of the unit vectors $\boldsymbol{i}, \boldsymbol{j}$, and $\boldsymbol{k}$, i.e.

$$\boldsymbol{a} = \alpha_x \boldsymbol{i} + \alpha_y \boldsymbol{j} + \alpha_z \boldsymbol{k}$$

The vectors $\boldsymbol{a}_x = \alpha_x \boldsymbol{i}$, $\boldsymbol{a}_y = \alpha_y \boldsymbol{j}$, and $\boldsymbol{a}_z = \alpha_z \boldsymbol{k}$ are called the *rectangular (vector) components* of the vector $\boldsymbol{a}$.

Clearly the following is true:

Proposition 4.1.2

$$\boldsymbol{a} = \boldsymbol{b} \text{ iff } \alpha_x = \beta_x, \alpha_y = \beta_y, \alpha_z = \beta_z.$$

The *magnitude* or the *norm* of a vector in general 3-space has already been mentioned several times. Now we define it formally for $\mathbf{R}^3$.

Definition 4.1.2

Let $\boldsymbol{a}$ be any vector in $\mathbf{R}^3$. We say that the ***magnitude, modulus*** or ***norm***, $|\boldsymbol{a}|$ of a vector $\boldsymbol{a}$ is a number $a \in \mathbf{R}$ representing the length of $\boldsymbol{a}$, i.e., if $\boldsymbol{a} = \overrightarrow{AB}$, then

$$|\boldsymbol{a}| = \left|\overrightarrow{AB}\right| = \left|\overline{AB}\right| = a.$$

Thus, for the Cartesian coordinate system we have:

Definition 4.1.3

Let $\boldsymbol{a} = \alpha_x \boldsymbol{i} + \alpha_y \boldsymbol{j} + \alpha_z \boldsymbol{k}$ be any vector in $\mathbf{R}^3$. The magnitude of $\boldsymbol{a}$ is the number

$$a = |\boldsymbol{a}| = \sqrt{\alpha_x^2 + \alpha_y^2 + \alpha_z^2}.$$

Example 4.1.1

Let's first consider a vector in $\mathbf{R}^2$ (Figure 4.3):

Applying Pythagoras' theorem to the triangle $\Delta O a_x A$, we see that the length a, i.e., the magnitude of $\boldsymbol{a}$, is

$$a = |\boldsymbol{a}| = \sqrt{\alpha_x^2 + \alpha_y^2}.$$

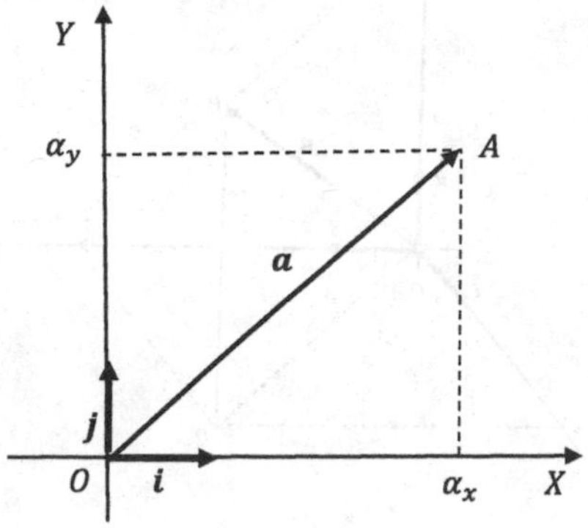

FIGURE 4.3

It is not difficult now to extend this to three dimensions (Figure 4.4):

$$a = |\boldsymbol{a}| = \sqrt{\alpha_x^2 + \alpha_y^2 + \alpha_z^2}\,. \qquad \blacksquare$$

Example 4.1.2
Let $\boldsymbol{a} = \alpha_x i + \alpha_y j + \alpha_z k$ be a vector in $\mathbf{R}^3$ (Figure 4.4).
Then

$$|\boldsymbol{a}| = \overline{OP},$$

with

$$\overline{OA} = \alpha_x,\, \overline{OB} = \alpha_y,\, \overline{QP} = \alpha_z.$$

By Pythagoras' theorem

$$\left(\overline{OP}\right)^2 = \left(\overline{OQ}\right)^2 + \left(\overline{QP}\right)^2$$

and

$$\left(\overline{OQ}\right)^2 = \left(\overline{OA}\right)^2 + \left(\overline{OB}\right)^2.$$

Thus

$$\begin{aligned}\left(\overline{OP}\right)^2 &= \left(\overline{OQ}\right)^2 + \left(\overline{QP}\right)^2 \\ &= \left(\overline{OA}\right)^2 + \left(\overline{OB}\right)^2 + \left(\overline{QP}\right)^2.\end{aligned}$$

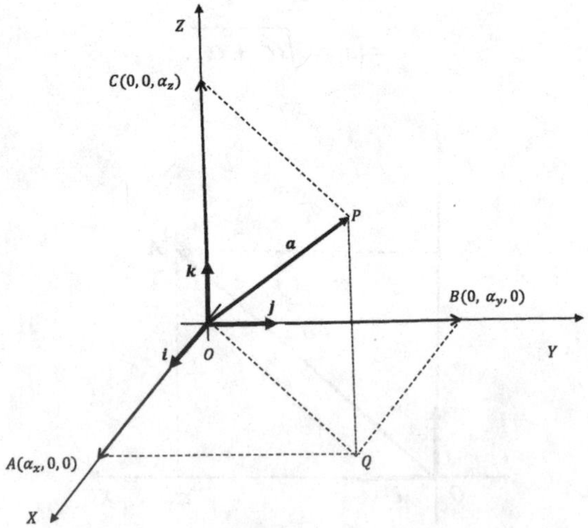

FIGURE 4.4

In other words,

$$|\boldsymbol{a}|^2 = \alpha_x^2 + \alpha_y^2 + \alpha_z^2.$$

So,

$$a = |\boldsymbol{a}| = \sqrt{\alpha_x^2 + \alpha_y^2 + \alpha_z^2}.$$

For example, if $\boldsymbol{a} = 2\boldsymbol{i} + 3\boldsymbol{j}$, then

$$a = |\boldsymbol{a}| = \sqrt{\alpha_x^2 + \alpha_y^2} = \sqrt{4+9} = \sqrt{13},$$

and if $\boldsymbol{a} = 2\boldsymbol{i} + 3\boldsymbol{j} + 4\boldsymbol{k}$, then

$$a = |\boldsymbol{a}| = \sqrt{\alpha_x^2 + \alpha_y^2 + \alpha_z^2} = \sqrt{4+9+16} = \sqrt{29}$$

■

Now it is obvious that

$$|\boldsymbol{i}| = \sqrt{1^2 + 0 + 0} = 1,$$

and, similarly, for $|\boldsymbol{j}|$ and $|\boldsymbol{k}|$. Of course,

$$|\boldsymbol{0}| = 0\boldsymbol{i} + 0\boldsymbol{j} + 0\boldsymbol{k} = 0.$$

Definition 4.1.4
A ***unit vector*** $\boldsymbol{a}_0$ is a vector whose magnitude is 1, i.e., if $\boldsymbol{a} \in \mathbf{R}^3$ is any vector such that $|\boldsymbol{a}| \neq 0$, then

$$\boldsymbol{a}_0 = \frac{\boldsymbol{a}}{a}$$

is a unit vector with the same direction as $\boldsymbol{a}$.

Indeed,

$$|\,\boldsymbol{a}_0\,| = \left|\frac{1}{a}\boldsymbol{a}\right| = \frac{1}{a}|\boldsymbol{a}| = \frac{1}{a}a = 1.$$

Exercise 4.1.2
Find the unit vectors $\boldsymbol{a}_0$, $\boldsymbol{b}_0$ and $\boldsymbol{c}_0$ if

(i) $\boldsymbol{a} = 2\boldsymbol{i} + 2\boldsymbol{j} - \boldsymbol{k}$ and $\boldsymbol{b} = 2\boldsymbol{i} + 2\boldsymbol{j} + 2\boldsymbol{k}$;

(ii) $\boldsymbol{a} = (1,2,1)$, $\boldsymbol{b} = (2,2,-1)$ and $\boldsymbol{c} = \left(\frac{3}{5}, 0, \frac{4}{5}\right)$.

Exercise 4.1.3
Show that, while $\boldsymbol{i}, \boldsymbol{j}$ and $\boldsymbol{k}$ are (obviously) non-coplanar, $\boldsymbol{i}, \boldsymbol{i}+\boldsymbol{j}$, and $\boldsymbol{i}-\boldsymbol{j}$ are coplanar.

Remark 4.1.1
We have said before that a vector $\boldsymbol{a}$ is *uniquely* represented once a triple $(\alpha_x, \alpha_y, \alpha_z)$ is given, meaning that we can determine its magnitude and its direction. At this point it is irrelevant whether the components of the vector are written in a row or in a column.[1]

Remark 4.1.2
As a reader might have anticipated by now, we can generalize the concept of the magnitude of vector to the n-dimensional case, and state:

If $x_1, x_2, \ldots, x_n \in \mathbf{R}$ are the coordinates of a vector $\boldsymbol{x} \in \mathbf{R}^n$, then by the magnitude of $\boldsymbol{x}$ we mean the number

$$|\boldsymbol{x}| = \sqrt{x_1^2 + x_2^2 + \ldots + x_n^2}.$$

Exercise 4.1.4
Let $\boldsymbol{x} = (x_1, x_2)$ and $\boldsymbol{y} = (y_1, y_2)$ be two vectors from $\mathbf{R}^2$. Show that if $x_1 y_2 - x_2 y_1 = 0$, then $\boldsymbol{x}$ and $\boldsymbol{y}$ are linearly dependent, and if $x_1 y_2 - x_2 y_1 \neq 0$ they are linearly independent.

Exercise 4.1.5
Let $\boldsymbol{a}, \boldsymbol{b} \in \mathbf{R}^3$. Show that

$$|\boldsymbol{a}-\boldsymbol{b}|^2 = |\boldsymbol{a}|^2 + |\boldsymbol{b}|^2 - 2(\alpha_x \beta_x + \alpha_y \beta_y + \alpha_z \beta_z).$$

4.2 Multiplication by a Scalar and Addition of Vectors in $\mathbf{R}^3$ Space

Definition 4.2.1 (Multiplication by a scalar)
Let $\boldsymbol{a} \in \mathbf{R}^3$ be any vector and $\lambda \in \mathbf{R}$ be any scalar. Then

$$\begin{aligned}\lambda \boldsymbol{a} &= \lambda\left(\alpha_x \boldsymbol{i} + \alpha_y \boldsymbol{j} + \alpha_z \boldsymbol{k}\right) \\ &= \lambda \alpha_x \boldsymbol{i} + \lambda \alpha_y \boldsymbol{j} + \lambda \alpha_z \boldsymbol{k} \\ &= \left(\lambda \alpha_x,\ \lambda \alpha_y,\ \lambda \alpha_z\right).\end{aligned}$$

In other words, the components of a scalar multiple of a vector are equal to the scalar multiple of the corresponding components of the vector.

In complete agreement with Definition 3.2.1 we have

(i) If $a \neq 0$, then $|a| > 0$;
(ii) $|\lambda a| = |\lambda||a|$;
(iii) For every $a \in \mathbf{R}^3$, $|-a| = |a|$.

Exercise 4.2.1
Let $a = i + 2j + 3k$. Show that

(i) $3a = 3i + 6j + 9k$;
(ii) $|-a| = |a| = \sqrt{14}$.

Definition 4.2.2 (Addition of vectors in $\mathbf{R}^3$)
If $a = (\alpha_x, \alpha_y, \alpha_z), b = (\beta_x, \beta_y, \beta_z) \in \mathbf{R}^3$ are any two vectors, with $\alpha_x, \alpha_y, \alpha_z, \beta_x, \beta_y, \beta_z \in \mathbf{R}$, then

$$\begin{aligned} a + b &= (\alpha_x i + \alpha_y j + \alpha_z k) + (\beta_x i + \beta_y j + \beta_z k) \\ &= (\alpha_x + \beta_x) i + (\alpha_y + \beta_y) j + (\alpha_z + \beta_z) k \\ &= (\alpha_x + \beta_x,\ \alpha_y + \beta_y,\ \alpha_z + \beta_z). \end{aligned}$$

In general, if $x_q = \alpha_q i + \beta_q j + \gamma_q k \in \mathbf{R}^3$, $q = 1, 2, \ldots, n$, are any n vectors, then

$$\begin{aligned} \sum_{q=1}^{n} x_q &= \left(\sum_{q=1}^{n} \alpha_q\right) i + \left(\sum_{q=1}^{n} \beta_q\right) j + \left(\sum_{q=1}^{n} \gamma_q\right) k \\ &= \left(\sum_{q=1}^{n} \alpha_q, \sum_{q=1}^{n} \beta_q, \sum_{q=1}^{n} \gamma_q\right). \end{aligned}$$

Since the addition of real numbers is a commutative operation, it is also obvious that

$$a + b = b + a.$$

It follows that $|a + b| = |b + a|$.

Proposition 4.2.1
The components of the sum (difference) of two vectors are equal to the sum (difference) of the corresponding components of the vectors.

Example 4.2.1
Let $a = i + 2j + 3k$ and $b = 4i - j + 2k$. Show that $a + b = b + a$.

Solution

$$\begin{aligned} a+b &= (i+2j+3k)+(4i-j+2k) \\ &= (1+4)i+(2-1)j+(3+2)k \\ &= (4+1)i+(-1+2)j+(2+3)k \\ &= (4i-j+2k)+(i+2j+3k) \\ &= b+a = 5i+j+5k. \end{aligned}$$

■

Exercise 4.2.2
Let $d = a-b+c$, with
$a=-i+2j+2k, b=3i-2j+k$, and $c=2i-4j-3k$. Find $|d|$.

As a simple exercise the reader should prove the following:

Proposition 4.2.2
The addition of vectors is an associative operation.

Exercise 4.2.3
Let $a=i+2j+3k, b=4i-j+2k$ and $c=5i+3j-k$. Show that

$$a+(b+c)=(a+b)+c.$$

Exercise 4.2.4
Let $a,b \in \mathbf{R}^3$ and $\lambda,\mu \in \mathbf{R}$. Show that

(i) $\lambda(a+b)=\lambda a+\lambda b$;
(ii) $\lambda a+\mu a=(\lambda+\mu)a$;
(iii) $\lambda(\mu a)=(\lambda\mu)a$.

Definition 4.2.3
Let the position vectors for points $A(\alpha_x,\alpha_y,\alpha_z)$ and $B(\beta_x,\beta_y,\beta_z)$ be

$$a=\alpha_x i+\alpha_y j+\alpha_z k,$$

$$b=\beta_x i+\beta_y \mathrm{j}+\beta_z k,$$

(Figure 4.5).

Then the distance $d(A,B)$ between points A and B is

$$\begin{aligned} d(A,B) &= |b-a| = \sqrt{(b-a)\times(b-a)} \\ &= \sqrt{(\beta_x-\alpha_x)^2+(\beta_y-\alpha_y)^2+(\beta_z-\alpha_z)^2}. \end{aligned}$$

■

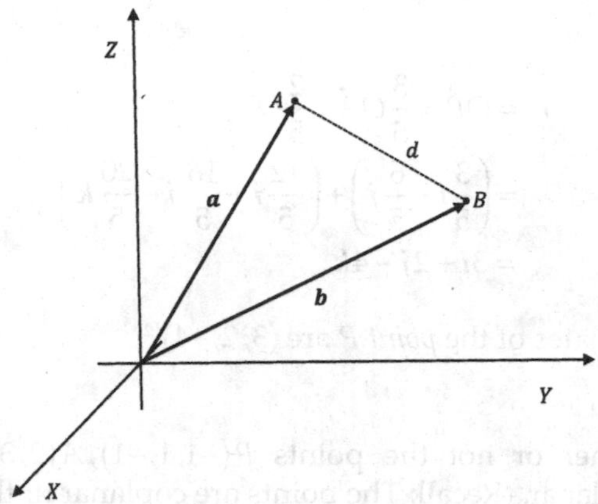

FIGURE 4.5

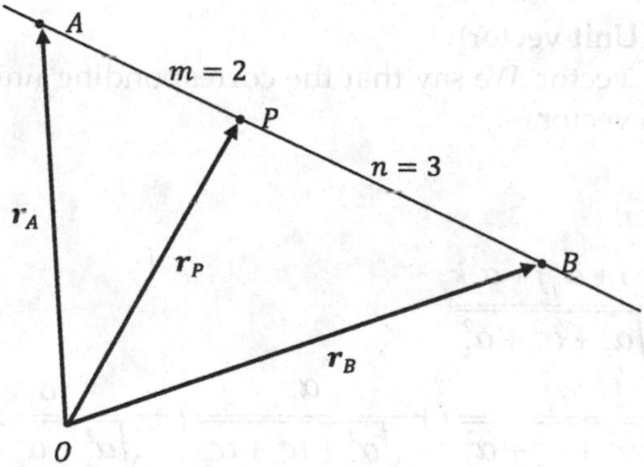

FIGURE 4.6

Exercise 4.2.5
Find the distance of the point $P(3,2,-4)$ from the origin $O(0,0,0)$.

Example 4.2.2
Determine the coordinates of the point which divides the line segment from $A(1,-2,0)$ to $B(6,8,-10)$ in the ratio 2 : 3.

Solution
Let point P be as in Figure 4.6,
and let

$$\boldsymbol{r}_A = \overrightarrow{OA} = \boldsymbol{i} - 2\boldsymbol{j} \quad \text{and} \quad \boldsymbol{r}_B = \overrightarrow{OB} = 6\boldsymbol{i} + 8\boldsymbol{j} - 10\boldsymbol{k}.$$

So

$$\begin{aligned}
r_P &= \overrightarrow{OP} = \frac{3}{5}\overrightarrow{OA} + \frac{2}{5}\overrightarrow{OB} \\
&= \left(\frac{3}{5}i - \frac{6}{5}j\right) + \left(\frac{12}{5}i + \frac{16}{5}j - \frac{20}{5}k\right) \\
&= 3i + 2j - 4k
\end{aligned}$$

Hence the coordinates of the *point* P are $(3,2,-4)$.[2] ∎

Exercise 4.2.6
Determine whether or not the points $P(-1,1,-1), A(2,3,0), B(0,1,1)$ and $C(-2,2,2)$ are coplanar. (Recall: The points are coplanar iff the vectors $\overrightarrow{PA}, \overrightarrow{PB}$ and $\overrightarrow{PC}$ are linearly dependent.)

Given the components of a vector $\boldsymbol{a} \in \mathbf{R}^3$ we restate Definition 4.1.4 as

Definition 4.2.4 (Unit vector)
Let $\boldsymbol{a} \in \mathbf{R}^3$ be any vector. We say that the corresponding unit vector $\boldsymbol{a}_0$ in the direction of $\boldsymbol{a}$ is a vector

$$\begin{aligned}
\boldsymbol{a}_0 &= \frac{1}{a}\boldsymbol{a}. \\
&= \frac{\alpha_x \boldsymbol{i} + \alpha_y \boldsymbol{j} + \alpha_z \boldsymbol{k}}{\sqrt{\alpha_x^2 + \alpha_y^2 + \alpha_z^2}} \\
&= \frac{\alpha_x}{\sqrt{\alpha_x^2 + \alpha_y^2 + \alpha_z^2}}\boldsymbol{i} + \frac{\alpha_y}{\sqrt{\alpha_x^2 + \alpha_y^2 + \alpha_z^2}}\boldsymbol{j} + \frac{\alpha_z}{\sqrt{\alpha_x^2 + \alpha_y^2 + \alpha_z^2}}\boldsymbol{k}.
\end{aligned}$$

Example 4.2.3
Let $\boldsymbol{a} = -2\boldsymbol{i} + 3\boldsymbol{j} + 5\boldsymbol{k}$. Show that

$$\boldsymbol{a}_0 = \left(\frac{-2}{\sqrt{38}}, \frac{3}{\sqrt{38}}, \frac{5}{\sqrt{38}}\right).$$

∎

Exercise 4.2.7
Show that $\boldsymbol{a}_0$ from Definition 4.2.4 and $\boldsymbol{a}_0$ from the previous example are indeed such that $|\boldsymbol{a}_0| = 1$.

Example 4.2.4
Find the magnitude of the position vector $\boldsymbol{r}_P$ of the point $P(3,-4,12)$ and the corresponding unit vector $\boldsymbol{r}_{P_0}$.

Solution

$$\boldsymbol{r}_P = \overrightarrow{OP} = 3\boldsymbol{i} - 4\boldsymbol{j} + 12\boldsymbol{k}.$$

So,

$$r_P = |\,\boldsymbol{r}_P\,| = \sqrt{9+16+144} = 13,$$

and

$$\boldsymbol{r}_{P_0} = \frac{\boldsymbol{r}_P}{|\,\boldsymbol{r}_P\,|} = \frac{\boldsymbol{r}_P}{r_P} = \frac{3}{13}\boldsymbol{i} - \frac{4}{13}\boldsymbol{j} + \frac{12}{13}\boldsymbol{k}.$$

■

Example 4.2.5
Let $\boldsymbol{a} = 3\boldsymbol{i} - \boldsymbol{j} - \boldsymbol{k}$ and $\boldsymbol{b} = 2\boldsymbol{i} + 2\boldsymbol{j} - 4\boldsymbol{k}$. Find $(\boldsymbol{a}+\boldsymbol{b})_0$.

Solution
Let

$$\boldsymbol{c} = \boldsymbol{a} + \boldsymbol{b} = (3\boldsymbol{i} - \boldsymbol{j} - \boldsymbol{k}) + (2\boldsymbol{i} + 2\boldsymbol{j} - 4\boldsymbol{k}) = 5\boldsymbol{i} + \boldsymbol{j} - 5\boldsymbol{k}$$

Then

$$c = |\boldsymbol{c}| = \sqrt{25+1+25} = \sqrt{51}.$$

Hence

$$(\boldsymbol{a}+\boldsymbol{b})_0 = \boldsymbol{c}_0 = \frac{5}{\sqrt{51}}\boldsymbol{i} + \frac{1}{\sqrt{51}}\boldsymbol{j} - \frac{5}{\sqrt{51}}\boldsymbol{k}.$$

■

Now that the concept of a vector as an "object" having magnitude and direction has been introduced, and the rules for the addition of such objects are clear to us, we ask: can we equally well define multiplication? Well, unlike addition, the concept of a vector does not contain in itself any indication as to how the product should be defined. So, in principle, one is free to choose a definition that best suits further mathematical and scientific application. Of course, the chosen definition has to be consistent with the rules introduced earlier. It turns out that, because of their particular nature, vectors can be multiplied in two different ways. We start with the scalar product.

4.3 Scalar (Dot) Product of Vectors

Definition 4.3.1 (Scalar or dot product of vectors)
Let $\mathbf{R}^3$ be the real vector space. Then the map

$$s:\mathbf{R}^3\times\mathbf{R}^3\to\mathbf{R}$$

called the ***scalar product*** or ***dot product*** is defined as follows:

For any two vectors $\boldsymbol{a},\boldsymbol{b}\in\mathbf{R}^3$, with $a=|\boldsymbol{a}|$, $b=|\boldsymbol{b}|$ and $\theta=\measuredangle(\boldsymbol{a},\boldsymbol{b})$ being the angle between $\boldsymbol{a}$ and $\boldsymbol{b}$,

$$\boldsymbol{a}\cdot\boldsymbol{b}=ab\cos(\boldsymbol{a},\boldsymbol{b})=ab\cos\theta,0\le\theta\le\pi.$$

This product has the following properties:

(i) $\boldsymbol{a}\cdot\boldsymbol{b}=\boldsymbol{b}\cdot\boldsymbol{a}$;
(ii) $\boldsymbol{a}\cdot(\boldsymbol{b}+\boldsymbol{c})=\boldsymbol{a}\cdot\boldsymbol{b}+\boldsymbol{a}\cdot\boldsymbol{c}$;
(iii) $\lambda(\boldsymbol{a}\cdot\boldsymbol{b})=(\lambda\boldsymbol{a})\cdot\boldsymbol{b}=\boldsymbol{a}(\lambda\boldsymbol{b})=(\boldsymbol{a}\cdot\boldsymbol{b})\lambda$;
(iv) $\boldsymbol{a}\cdot\boldsymbol{b}>0 \text{ iff } 0\le\theta<\frac{\pi}{2}$ and $\boldsymbol{a},\boldsymbol{b}\ne0$;
(v) $\boldsymbol{a}\cdot\boldsymbol{b}<0 \text{ iff } \frac{\pi}{2}<\theta\le\pi$ and $\boldsymbol{a},\boldsymbol{b}\ne0$; and
(vi) $\boldsymbol{a}\cdot\boldsymbol{b}=0 \text{ iff } \boldsymbol{a}\perp\boldsymbol{b}$

Proposition 4.3.1
The scalar product of two vectors $\boldsymbol{a}=\alpha_x\boldsymbol{i}+\alpha_y\boldsymbol{j}+\alpha_z\boldsymbol{k},\boldsymbol{b}=\beta_x\boldsymbol{i}+\beta_y+\beta_z\in\mathbf{R}^3$ is equal to the sum of the products of their corresponding components, i.e.

$$\boldsymbol{a}\cdot\boldsymbol{b}=\alpha_x\beta_x+\alpha_y\beta_y+\alpha_z\beta_z.$$

Proof

$$\begin{aligned}\boldsymbol{a}\cdot\boldsymbol{b}&=(\alpha_x\boldsymbol{i}+\alpha_y\boldsymbol{j}+\alpha_z\boldsymbol{k})\cdot(\beta_x\boldsymbol{i}+\beta_y\boldsymbol{j}+\beta_z\boldsymbol{k})\\
&=\alpha_x\beta_x(\boldsymbol{i}\cdot\boldsymbol{i})+\alpha_x\beta_y(\boldsymbol{i}\cdot\boldsymbol{j})+\alpha_x\beta_z(\boldsymbol{i}\cdot\boldsymbol{k})+\\
&+\alpha_y\beta_x(\boldsymbol{j}\cdot\boldsymbol{i})+\alpha_y\beta_y(\boldsymbol{j}\cdot\boldsymbol{j})+\alpha_y\beta_z(\boldsymbol{j}\cdot\boldsymbol{k})+\\
&+\alpha_z\beta_x(\boldsymbol{k}\cdot\boldsymbol{i})+\alpha_z\beta_y(\boldsymbol{k}\cdot\boldsymbol{j})+\alpha_z\beta_z(\boldsymbol{k}\cdot\boldsymbol{k})\\
&=\alpha_x\beta_x\cdot1+\alpha_x\beta_y\cdot0+\alpha_x\beta_z\cdot0+\\
&+\alpha_y\beta_x\cdot0+\alpha_y\beta_y\cdot1+\alpha_y\beta_z\cdot0+\\
&+\alpha_z\beta_x\cdot0+\alpha_z\beta_y\cdot0+\alpha_z\beta_z\cdot1\\
&=\alpha_x\beta_x+\alpha_y\beta_y+\alpha_z\beta_z.\end{aligned}$$

■

Corollary 4.3.1
If $\boldsymbol{a} = \alpha_x \boldsymbol{i} + \alpha_y \boldsymbol{j} + \alpha_z \boldsymbol{k} \in \mathbf{R}^3$ is any vector, then

$$\begin{aligned} a^2 &= \boldsymbol{a} \cdot \boldsymbol{a} = \alpha_x^2 + \alpha_y^2 + \alpha_z^2, \\ &= a \cdot a \cdot \cos 0 = a^2, \\ &= \left|a^2\right| = a^2 \geq 0. \end{aligned}$$

Let's emphasize again,

$$a = \sqrt{\boldsymbol{a} \cdot \boldsymbol{a}} = \sqrt{\alpha_x^2 + \alpha_y^2 + \alpha_z^2}.$$

Let's prove property (vi) of the scalar product, i.e., let's prove the following:

Proposition 4.3.2
Let $\boldsymbol{a} \neq \mathbf{0}, \boldsymbol{b} \neq \mathbf{0}$ be any two vectors from $\mathbf{R}^3$. Then they are perpendicular iff

$$\boldsymbol{a} \cdot \boldsymbol{b} = 0.$$

Proof

Let $\boldsymbol{a} \perp \boldsymbol{b}$. Then $\measuredangle(\boldsymbol{a}, \boldsymbol{b}) = \frac{\pi}{2}$, and we have

$$\boldsymbol{a} \cdot \boldsymbol{b} = ab \cos \frac{\pi}{2} = 0$$

Conversely, if $\boldsymbol{a} \cdot \boldsymbol{b} = ab \cos(\boldsymbol{a}, \boldsymbol{b}) = 0$, while $\boldsymbol{a} \neq \mathbf{0}, \boldsymbol{b} \neq \mathbf{0}$, then $\cos(\boldsymbol{a}, \boldsymbol{b}) = 0$. But this is possible only if $\measuredangle(\boldsymbol{a}, \boldsymbol{b}) = \frac{\pi}{2}$. ■

Corollary 4.3.2
Vectors $\boldsymbol{a} = \alpha_x \boldsymbol{i} + \alpha_y \boldsymbol{j} + \alpha_z \boldsymbol{k}$ and $\boldsymbol{b} = \beta_x \boldsymbol{i} + \beta_y \boldsymbol{j} + \beta_z \boldsymbol{k}$ are perpendicular iff

$$\alpha_x \beta_x + \alpha_y \beta_y + \alpha_z \beta_z = 0.$$

Exercise 4.3.1
Let $\boldsymbol{a} = (1, 2, 3)$ and $\boldsymbol{b} = (2, 3, 5)$. Find $\boldsymbol{a} \cdot \boldsymbol{b}$.

Exercise 4.3.2
Prove that $\boldsymbol{a} \cdot \boldsymbol{a} = 0$ iff $\boldsymbol{a} = \mathbf{0}$.

Exercise 4.3.3
Show that for any $\boldsymbol{a} = (\alpha_x, \alpha_y, \alpha_z) \in \boldsymbol{R}^3$,

(i) $a \cdot i = \alpha_x$;
(ii) $a \cdot j = \alpha_y$; and
(iii) $a \cdot k = \alpha_z$.

Example 4.3.1
Show that

(i) $i \cdot i = j \cdot j = k \cdot k = 1$; and
(ii) $i \cdot j = j \cdot k = k \cdot i = 0$.

Solution
(i) From Definition 4.3.1,

$$i \cdot i = 1 \cdot 1 \cdot \cos 0 = 1 \cdot 1 = 1.$$

Similarly for $j \cdot j$ and $k \cdot k$.

(ii) $i \cdot j = 1 \cdot 1 \cdot \cos \frac{\pi}{2} = 1 \cdot 0 = 0$.

Similarly for $j \cdot k$ and $k \cdot i$. ■

Exercise 4.3.4
Prove that if $a \cdot i = a \cdot j = a \cdot k = 0$, then $a = \mathbf{0}$.

Remark 4.3.1
It is important to note that the "cancelation law" *does not* hold for vectors. Consider three vectors a, b and c such that $a \neq b \neq c$, and suppose that $a \cdot b = a \cdot c$. Then

$$a \cdot b - a \cdot c = 0.$$

So, we have

$$a \cdot (b - c) = 0,$$

which means that a is perpendicular to $b - c$ while $b \neq c$.

Exercise 4.3.5
Show that a set of any two (three) vectors such that $a \cdot b = \mathbf{0}$ (or $a \cdot b = b \cdot c = c \cdot a = \mathbf{0}$) constitute an orthogonal basis for a 2- (3-)dimensional vector space.

Example 4.3.2
Let's verify (ii) from Definition 4.3.1:

$$
\begin{aligned}
a\cdot(b+c) &= \left(\alpha_x i+\alpha_y j+\alpha_z k\right)\left[\left(\beta_x i+\beta_y j+\beta_z k\right)+\left(\gamma_x i+\gamma_y j+\gamma_z k\right)\right] \\
&= \left(\alpha_x i+\alpha_y j+\alpha_z k\right)\cdot\left[(\beta_x+\gamma_x)i+\left(\beta_y+\gamma_y\right)j+(\beta_z+\gamma_z)k\right] \\
&= \alpha_x(\beta_x+\gamma_x)+\alpha_y\left(\beta_y+\gamma_y\right)+\alpha_z(\beta_z+\gamma_z) \\
&= \alpha_x\beta_x+\alpha_x\gamma_x+\alpha_y\beta_y+\alpha_y\gamma_y+\alpha_z\beta_z+\alpha_z\gamma_z \\
&= \left(\alpha_x\beta_x+\alpha_y\beta_y+\alpha_z\beta_z\right)+\left(\alpha_x\gamma_x+\alpha_y\gamma_y+\alpha_z\gamma_z\right) \\
&= a\cdot b+a\cdot c.
\end{aligned}
$$

■

Exercise 4.3.6
Let $a=i+2j+3k, b=-i+k$, and $c=i-2j+5k$. Find $(b-a)\cdot(c-a)$.

Proposition 4.3.3
Let $a,b\in\mathbf{R}^3$ be any two vectors. Then,

$$|a+b|=|a-b| \text{ iff } a\cdot b=0.$$

Proof

$$
\begin{aligned}
|a+b|=|a-b| &\Leftrightarrow |a+b|^2=|a-b|^2 \\
&\Leftrightarrow a^2+2a\cdot b+b^2=a^2-2a\cdot b+b^2 \\
&\Leftrightarrow 4a\cdot b=0 \\
&\Leftrightarrow a\cdot b=0.
\end{aligned}
$$

■

Definition 4.3.2 (Direction cosines)
Consider a non-zero vector $a\in\mathbf{R}^3$ and the angles $\theta=\measuredangle(a,i), \phi=\measuredangle(a,j), \psi=\measuredangle(a,k)$ formed with the vector a and the unit vectors i, j and k (Figure 4.7).

We say that

$$\cos\theta=\frac{i\cdot a}{a},$$

$$\cos\phi=\frac{j\cdot a}{a},$$

$$\cos\psi=\frac{k\cdot a}{a},$$

are ***direction cosines***. In other words, the direction of a vector a is uniquely determined by the angles θ, ϕ, and ψ.

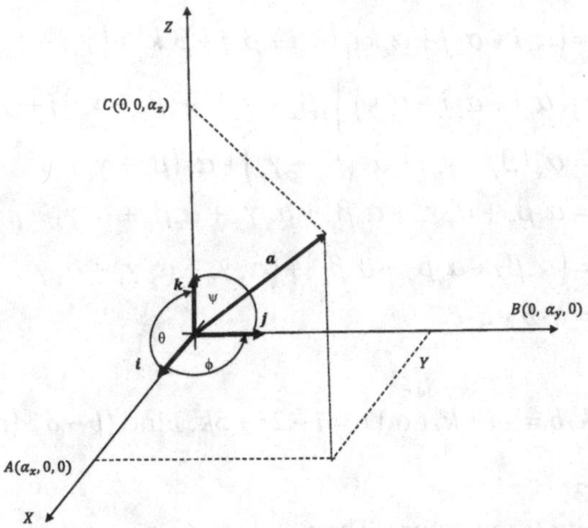

FIGURE 4.7

Example 4.3.3
If a vector $\boldsymbol{a}$ is given via its components, i.e. $\boldsymbol{a}=(\alpha_x,\alpha_y,\alpha_z)$, we can easily determine its direction in $\mathbf{R}^3$; that is, we can find out what angles vector $\boldsymbol{a}$ makes with the coordinate axes X, Y, and Z in the following way (Figure 4.7).

So, with $\theta=\sphericalangle(\boldsymbol{a},\boldsymbol{i})$, $\varphi=\sphericalangle(\boldsymbol{a},\boldsymbol{j})$ and $\psi=\sphericalangle(\boldsymbol{a},\boldsymbol{k})$ we have

$$\begin{aligned}\boldsymbol{i}\cdot\boldsymbol{a}&=1\cdot a\cdot\cos\vartheta=a\cos\vartheta\\&=\boldsymbol{i}\cdot(\alpha_x\boldsymbol{i}+\alpha_y\boldsymbol{j}+\alpha_z\boldsymbol{k})=\alpha_x\end{aligned}$$

Thus

$$\cos\vartheta=\frac{\alpha_x}{a}=\frac{\alpha_x}{\sqrt{\alpha_x^2+\alpha_y^2+\alpha_z^2}}. \tag{4.1}$$

Similarly,

$$\begin{aligned}\boldsymbol{j}\cdot\boldsymbol{a}&=1\cdot a\cdot\cos\varphi=a\cos\varphi\\&=\boldsymbol{i}\cdot(\alpha_x\boldsymbol{i}+\alpha_y\boldsymbol{j}+\alpha_z\boldsymbol{k})=\alpha_y;\end{aligned}$$

$$\cos\varphi=\frac{\alpha_y}{a}=\frac{\alpha_y}{\sqrt{\alpha_x^2+\alpha_y^2+\alpha_z^2}}. \tag{4.2}$$

And finally,

$$\begin{aligned}\boldsymbol{k}\cdot\boldsymbol{a}&=1\cdot a\cdot\cos\psi=a\cos\psi\\&=\boldsymbol{k}\cdot(\alpha_x\boldsymbol{i}+\alpha_y\boldsymbol{j}+\alpha_z\boldsymbol{k})=\alpha_z\end{aligned}$$

$$\cos\psi=\frac{\alpha_z}{a}=\frac{\alpha_z}{\sqrt{\alpha_x^2+\alpha_y^2+\alpha_z^2}}. \tag{4.3}$$

$(1),(2)$ and (3) are the cosines of the angles that vector $\boldsymbol{a}$ makes with the coordinate axes X, Y and Z. ■

The way to find the angle between *any* two vectors is illustrated by the following two examples.

Example 4.3.4
Find the angle between $\boldsymbol{a} = \alpha_x \boldsymbol{i} + \alpha_y \boldsymbol{j} + \alpha_z \boldsymbol{k}$ and $\boldsymbol{b} = \beta_x \boldsymbol{i} + \beta_y \boldsymbol{j} + \beta_z \boldsymbol{k}$.

Solution
From

$$\boldsymbol{a} \cdot \boldsymbol{b} = ab \cos\theta$$

we get

$$\cos\theta = \frac{\boldsymbol{a} \cdot \boldsymbol{b}}{ab} = \frac{\alpha_x \beta_x + \alpha_y \beta_y + \alpha_z \beta}{\sqrt{\alpha_x^2 + \alpha_y^2 + \alpha_z^2}\sqrt{\beta_x^2 + \beta_y^2 + \beta_z^2}}$$

Hence

$$\theta = \cos^{-1}\left(\frac{\alpha_x \beta_x + \alpha_y \beta_y + \alpha_z \beta}{\sqrt{\alpha_x^2 + \alpha_y^2 + \alpha_z^2}\sqrt{\beta_x^2 + \beta_y^2 + \beta_z^2}}\right).$$

■

Example 4.3.5
Let $\boldsymbol{a} = -2\boldsymbol{i} + 3\boldsymbol{j} + \boldsymbol{k}$ and $\boldsymbol{b} = \boldsymbol{i} + 2\boldsymbol{j} + 2\boldsymbol{k}$. Find the angle θ between $\boldsymbol{a}$ and $\boldsymbol{b}$.

Solution
Since

$$\boldsymbol{a} \cdot \boldsymbol{b} = (-2\boldsymbol{i} + 3\boldsymbol{j} + \boldsymbol{k}) \cdot (\boldsymbol{i} + 2\boldsymbol{j} + 2\boldsymbol{k}) = -2 + 6 + 2 = 6$$

and

$$a = \sqrt{4+9+1} = \sqrt{14} \text{ and } b = \sqrt{1+4+4} = 3,$$

$$\cos\theta = \frac{\boldsymbol{a} \cdot \boldsymbol{b}}{ab} = \frac{6}{3\sqrt{14}} = \frac{2}{\sqrt{14}}.$$

Hence

$$\theta = \cos^{-1}\frac{2}{\sqrt{14}}.$$

■

Exercise 4.3.7
Let $a = i - 2j + 4k$, $b = 2i + j$ and $c = -i + j - 3k$. Find

$$\theta = \sphericalangle(a, a + 2b + 3c).$$

Exercise 4.3.8
Show that the angle between $a = 3i - 2j + 6k$ and $b = -3i - 5j + 8k$ is 45°.

Example 4.3.6
Let $a \in \mathbf{R}^3$ be any vector such that $\theta = \sphericalangle(a, i), \phi = \sphericalangle(a, j)$ and $\psi = \sphericalangle(a, k)$. Prove that

$$cos^2\theta + cos^2\phi + cos^2\psi = 1.$$

Proof

$$cos^2\theta + cos^2\phi + cos^2\psi = \left(\frac{\alpha_x}{a}\right)^2 + \left(\frac{\alpha_y}{a}\right)^2 + \left(\frac{\alpha_z}{a}\right)^2$$

$$= \frac{a_x^2}{\alpha_x^2 + \alpha_y^2 + \alpha_z^2} + \frac{\alpha_y^2}{\alpha_x^2 + \alpha_y^2 + \alpha_z^2} + \frac{\alpha_z^2}{\alpha_x^2 + \alpha_y^2 + \alpha_z^2} = 1.$$

Notice (cf. Definition 4.2.4), that for any $a \in \mathbf{R}^3$ we can now write

$$a_0 = \cos\theta i + \cos\phi j + \cos\psi k.$$

■

Example 4.3.7
Show that if $a \in \mathbf{R}^3$ is any vector, then

$$a = (a \cdot i)i + (a \cdot j)j + (a \cdot k)k.$$

Solution
Let $a = \alpha_x i + \alpha_y j + \alpha_z k$. Then

$$\begin{aligned}(a \cdot i)i &= (\alpha_x i \cdot i + \alpha_y j \cdot i + \alpha_z k \cdot i)i \\ &= \alpha_x i.\end{aligned}$$

$$\begin{aligned}(a \cdot j)j &= (\alpha_x i \cdot j + \alpha_y j \cdot j + \alpha_z k \cdot j)j \\ &= \alpha_y j.\end{aligned}$$

$$\begin{aligned}(a \cdot k)k &= (\alpha_x i \cdot k + \alpha_y j \cdot k + \alpha_z k \cdot k)k \\ &= \alpha_z k.\end{aligned}$$

So, indeed,

$$a = (a \cdot i)i + (a \cdot j)j + (a \cdot k)k \qquad \blacksquare$$

Now the following proposition, mentioned before in a different context, sounds rather obvious.

Proposition 4.3.4
Any vector from $\mathbf{R}^3$ is uniquely determined by its modulus and direction.

Definition 4.3.3
Let $a \neq 0$ be any vector from $\mathbf{R}^3$. Then for any other vector $x \in \mathbf{R}^3$ we define the (*scalar*) ***projection*** of x in the direction of a (Figure 4.8), as the scalar

$$x_a = x \cdot \cos\theta, \quad \theta = \measuredangle(a, x)$$

Example 4.3.8
Find the projection of $a = 3i - j - 2k$ on $b = i + 2j - 3k$.

Solution
Since

$$a \cdot b = ab\cos(a,b),$$

$$a\cos(a,b) = \frac{a \cdot b}{b}.$$

Therefore,

$$a\cos(a,b) = \frac{3 \cdot 1 + (-1) \cdot 2 + (-2) \cdot (-3)}{\sqrt{1^2 + 2^2 + (-3)^2}} = \frac{7}{\sqrt{14}}. \qquad \blacksquare$$

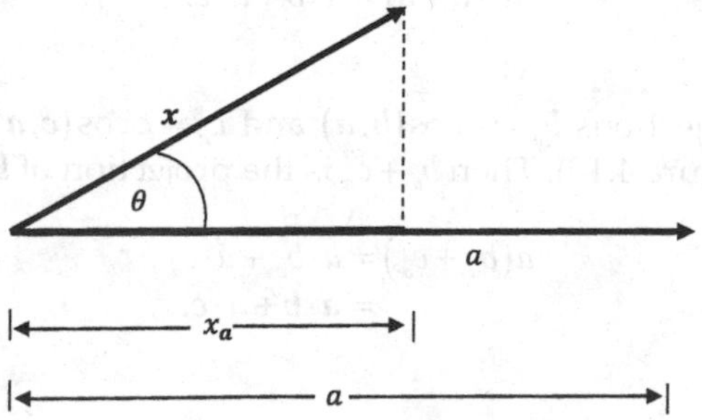

FIGURE 4.8

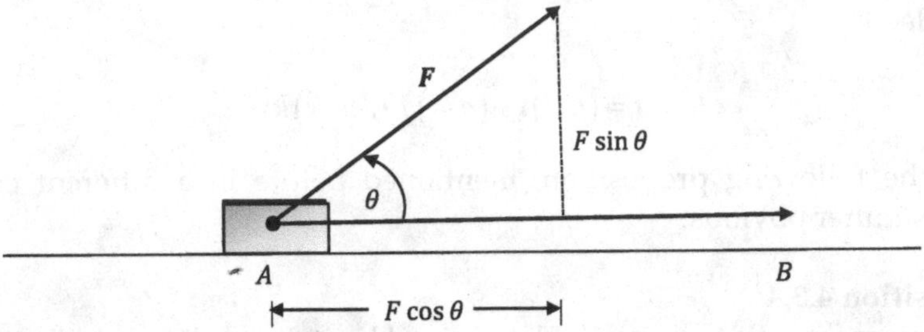

FIGURE 4.9

Example 4.3.9
A typical example of a scalar product from physics: the work W done by a force $\boldsymbol{F}$ acting on an object in moving it through a distance $|\boldsymbol{d}| = |\overrightarrow{AB}|$ is given by

$$W = \boldsymbol{F} \cdot \boldsymbol{d} = Fd\cos\theta$$

where θ is the angle between $\boldsymbol{F}$ and $\boldsymbol{d}$, and $F\cos\theta$ is the projection of force $\boldsymbol{F}$ in the direction of $\boldsymbol{d}$ (Figure 4.9). ■

Exercise 4.3.9
Show that

$$(x+y)_a = x_a + y_a.$$

Exercise 4.3.10
Show that, given two vectors $\boldsymbol{a}, \boldsymbol{b} \in \mathbf{R}^3$,

$$\boldsymbol{a} \cdot \boldsymbol{b} = a_b b = a b_a.$$

Example 4.3.10
Using the concepts of vector projections, let's show that

$$\boldsymbol{a} \cdot (\boldsymbol{b} + \boldsymbol{c}) = \boldsymbol{a} \cdot \boldsymbol{b} + \boldsymbol{a} \cdot \boldsymbol{c}.$$

Solution
Consider the projections $b_{\boldsymbol{a}} = b\cos(\boldsymbol{b}, \boldsymbol{a})$ and $c_{\boldsymbol{a}} = c\cos(\boldsymbol{c}, \boldsymbol{a})$ of the vectors $\boldsymbol{b}$ and $\boldsymbol{c}$ along $\boldsymbol{a}$ (Figure 4.10). Then $b_{\boldsymbol{a}} + c_{\boldsymbol{a}}$ is the projection of $\boldsymbol{b} + \boldsymbol{c}$ along $\boldsymbol{a}$. So,

$$\begin{aligned} a(b_{\boldsymbol{a}} + c_{\boldsymbol{a}}) &= a \cdot b_{\boldsymbol{a}} + a \cdot c_{\boldsymbol{a}} \\ &= \boldsymbol{a} \cdot \boldsymbol{b} + \boldsymbol{a} \cdot \boldsymbol{c}. \end{aligned}$$

■

Exercise 4.3.11
Let $\boldsymbol{a} = \boldsymbol{i} + 2\boldsymbol{j} + 3\boldsymbol{k}, \boldsymbol{b} = -\boldsymbol{i} + \boldsymbol{k}$, and $\boldsymbol{c} = \boldsymbol{i} - 2\boldsymbol{j} + 5\boldsymbol{k}$. Find $(\boldsymbol{b} - \boldsymbol{a}) \cdot (\boldsymbol{c} - \boldsymbol{a})$.

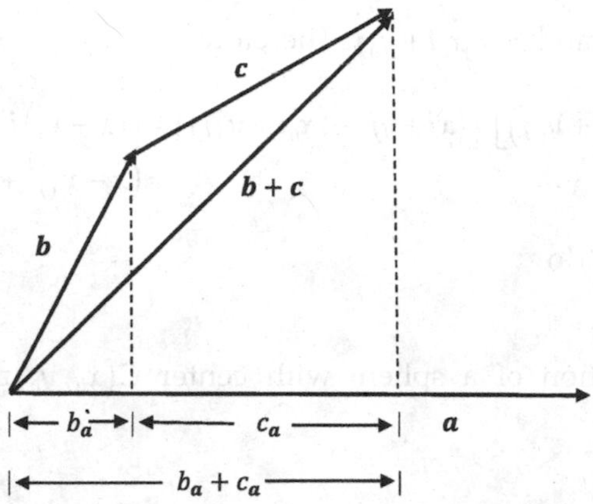

FIGURE 4.10

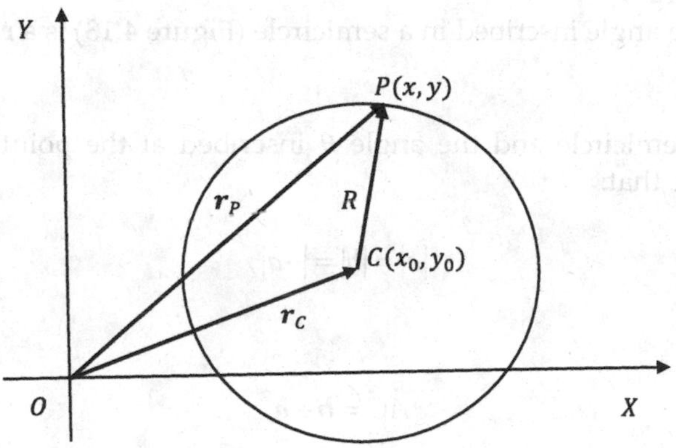

FIGURE 4.11

Example 4.3.11
As another example of the application of the scalar product of vectors, let's derive the equation of a circle with center $C(x_0, y_0)$ and radius R (Figure 4.11).

If $P(x, y)$ is any point on the circle, then $\left|\overrightarrow{CP}\right| = R$, and therefore $\overrightarrow{CP} \cdot \overrightarrow{CP} = R^2$. From Figure 4.11 we see that

$$\overrightarrow{CP} = \overrightarrow{OP} - \overrightarrow{OC} = \boldsymbol{r}_P - \boldsymbol{r}_C.$$

So we have

$$(\boldsymbol{r}_P - \boldsymbol{r}_C) \cdot (\boldsymbol{r}_P - \boldsymbol{r}_C) = R^2,$$

where $r_P = xi + yj$ and $r_C = x_0 i + y_0 j$. Therefore

$$[(xi+yj)-(x_0 i+y_0 j)]\cdot[(xi+yj)-(x_0 i+y_0 j)] = [(x-x_0)i+(y-y_0)j]^2$$
$$= (x-x_0)^2+(y-y_0)^2 = R^2. \quad \blacksquare$$

Similarly, one can do

Exercise 4.3.12
Derive the equation of a sphere with center $C(x_0, y_0, z_0)$ and radius R (Figure 4.12).

Exercise 4.3.13
Find the equation of the sphere with center at $C(3,-2,9)$ and radius $R = 5$.

Example 4.3.12
Show that the angle inscribed in a semicircle (Figure 4.13) is a right angle.

Solution
Consider a semicircle and the angle θ inscribed at the point C. From the figure, we see that:

$$|a| = |b| = |-a|,$$

and

$$\overrightarrow{AC} = b + a,$$

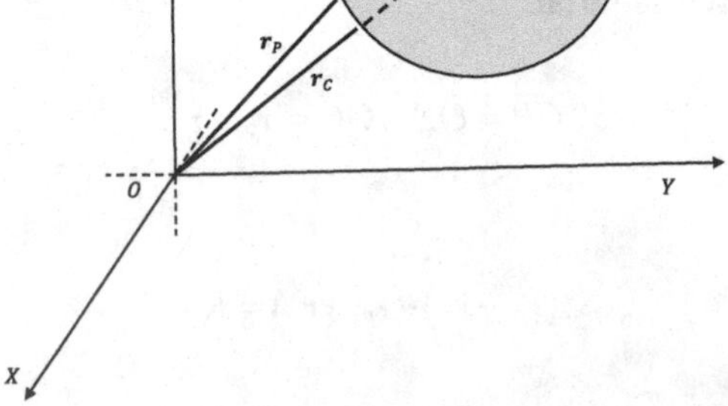

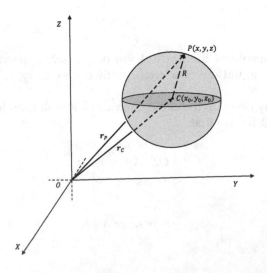

FIGURE 4.12

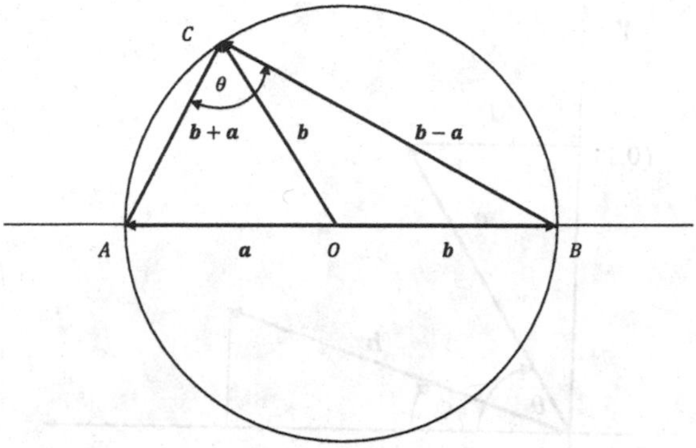

FIGURE 4.13

$$\overrightarrow{BC} = \boldsymbol{b} - \boldsymbol{a}$$

Then

$$\begin{aligned}\overrightarrow{AC} \cdot \overrightarrow{BC} &= (\boldsymbol{b}+\boldsymbol{a}) \cdot (\boldsymbol{b}-\boldsymbol{a}) \\ &= \boldsymbol{b} \cdot \boldsymbol{b} - \boldsymbol{a} \cdot \boldsymbol{a} \\ &= b^2 - a^2 \\ &= 0.\end{aligned}$$

Thus $\overrightarrow{AC}$ is perpendicular to $\overrightarrow{BC}$, i.e., the angle θ between them is a right angle. ■

Example 4.3.13

Let's prove the well-known trigonometric identity

$$\cos(\theta - \phi) = \cos\theta \sin\phi + \sin\theta \sin\phi.$$

Consider vectors $\boldsymbol{a}$ and $\boldsymbol{b}$ in a Cartesian coordinate system such that $a = b = 1$ (Figure 4.14).

Then

$$\boldsymbol{a} = \cos\theta \boldsymbol{i} + \sin\theta \boldsymbol{j},$$

and

$$\boldsymbol{b} = \cos\phi \boldsymbol{i} + \sin\phi \boldsymbol{j}.$$

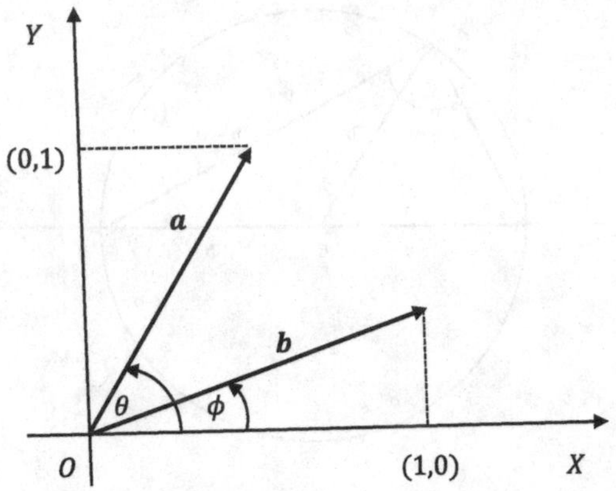

FIGURE 4.14

So

$$\begin{aligned}a\cdot b &= (\cos\theta i+\sin\theta j)\cdot(\cos\phi i+\sin\phi j)\\ &= ab\cos(a,b)\\ &= ab\cos(\theta-\phi)\\ &= 1\cdot 1\cdot\cos(\theta-\phi)\\ &= \cos\theta\cos\phi+\sin\theta\sin\phi.\end{aligned}$$

Thus,

$$\cos(\theta-\phi)=\cos\theta\sin\phi+\sin\theta\sin\phi.$$ ■

Example 4.3.14
For the next proposition we will need the cosine law.

Consider a triangle ΔABC (Figure 4.15):

From the figure we see that $\boldsymbol{b}+\boldsymbol{c}=\boldsymbol{a}$, or $\boldsymbol{c}=\boldsymbol{a}-\boldsymbol{b}$. Therefore,

$$c^2=\boldsymbol{c}\cdot\boldsymbol{c}=(\boldsymbol{a}-\boldsymbol{b})\cdot(\boldsymbol{a}-\boldsymbol{b})=\ \boldsymbol{a}\cdot\boldsymbol{a}+\boldsymbol{b}\cdot\boldsymbol{b}-2\boldsymbol{a}\cdot\boldsymbol{b}.$$

Thus,

$$c^2=a^2+b^2-2ab\cos\theta,$$

which is known as the cosine law. ■

Proposition 4.3.5
For any non-collinear $\boldsymbol{a},\boldsymbol{b}\in\mathbf{R}^3$,

(i) $(\boldsymbol{a}+\boldsymbol{b})^2=\boldsymbol{a}^2+\boldsymbol{b}^2+2\boldsymbol{a}\cdot\boldsymbol{b}=a^2+b^2+2ab\cos\theta$
(ii) $(\boldsymbol{a}-\boldsymbol{b})^2=\boldsymbol{a}^2+\boldsymbol{b}^2-2\boldsymbol{a}\cdot\boldsymbol{b}=a^2+b^2-2ab\cos\theta$

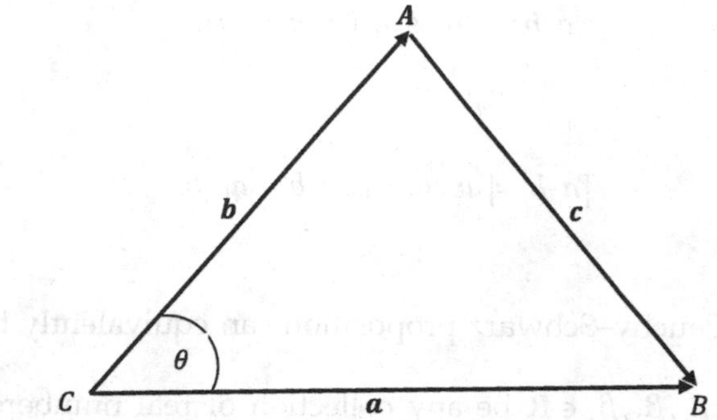

FIGURE 4.15

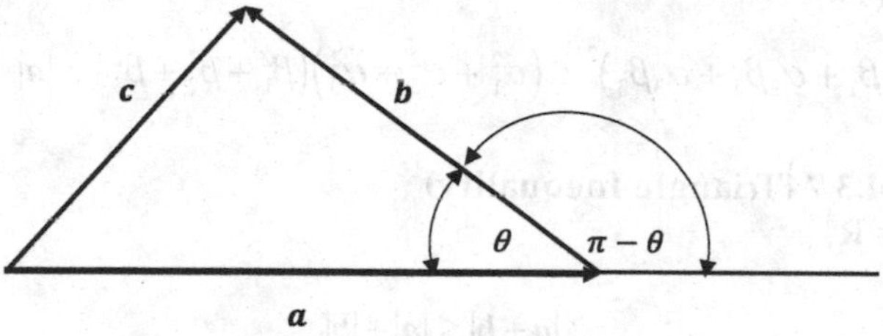

FIGURE 4.16

Proof
Let $\boldsymbol{a}+\boldsymbol{b}=\boldsymbol{c}$ where $\boldsymbol{a}$ and $\boldsymbol{b}$ are non-collinear (Figure 4.16).
Then

$$c^2 = (\boldsymbol{a}+\boldsymbol{b})^2.$$

Recalling the cosine theorem, we have

$$\begin{aligned} c^2 = c^2 &= a^2 + b^2 + 2ab\cos(\pi-\theta) \\ &= a^2 + b^2 - 2ab\cos\theta \\ &= \boldsymbol{a}^2 + \boldsymbol{b}^2 + 2\boldsymbol{a}\ \boldsymbol{b}. \end{aligned}$$

The proof for (ii) is analogous. ∎

Proposition 4.3.6 (Cauchy–Schwarz Inequality)
For any $\boldsymbol{a}, \boldsymbol{b} \in \mathbf{R}^3$

$$|\boldsymbol{a}\cdot\boldsymbol{b}| \le |\boldsymbol{a}|\cdot|\boldsymbol{b}|.$$

Proof
If $\boldsymbol{a}=\mathbf{0}$ or $\boldsymbol{b}=\mathbf{0}$, both sides of the inequality are immediately satisfied. So, let's assume that $\boldsymbol{a}\neq\mathbf{0}$ and $\boldsymbol{b}\neq\mathbf{0}$. Then

$$a \cdot b = ab\cos\theta,\ \theta = \measuredangle(a,b).$$

Since $|\cos\theta| \leq 1$,

$$|a \cdot b| = |ab\cos\theta| \leq a \cdot b = |a| \cdot |b|$$ ■

Exercise 4.3.14

Show that the Cauchy–Schwarz proposition can equivalently be proved as follows:

Let $\alpha_1, \alpha_2, \alpha_3, \beta_1, \beta_2, \beta_3 \in \mathbf{R}$ be any collection of real numbers considered as the coordinates of two vectors, a and b, respectively. Then the Cauchy–Schwarz proposition holds, i.e.,

$$(\alpha_1\beta_1 + \alpha_2\beta_2 + \alpha_3\beta_3)^2 \leq (\alpha_1^2 + \alpha_2^2 + \alpha_3^2)(\beta_1^2 + \beta_2^2 + \beta_3^2) = |a| \cdot |b|.$$

Proposition 4.3.7 (Triangle Inequality)

For any $a, b \in \mathbf{R}^3$

$$|a + b| \leq |a| + |b|$$

Proof

Since both sides of the inequality are non-negative, it suffices to prove the equivalent inequality

$$(|a + b|)^2 \leq (|a| + |b|)^2 \quad (*)$$

On the left-hand side of (*) we have

$$(a + b)(a + b) = a \cdot a + 2a \cdot b + b \cdot b,$$

while the right-hand side is

$$(|a|)^2 + 2|a||b| + (|b|)^2 = a^2 + 2ab + b^2.$$

Recalling Proposition 4.3.6 we conclude that $|a + b| \leq |a| + |b|$, as claimed. ■

Example 4.3.15

Prove that if the diagonals of a parallelogram are perpendicular, then the parallelogram is a rhombus.

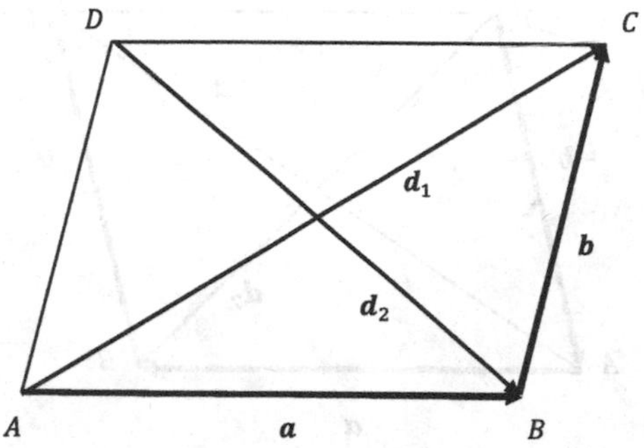

FIGURE 4.17

Proof

Let $ABCD$ be a parallelogram such that $\boldsymbol{a}$ and $\boldsymbol{b}$ represent sides AB and BC, respectively, i.e. $\boldsymbol{a} = \overrightarrow{AB}$ and $\boldsymbol{b} = \overrightarrow{BC}$ (Figure 4.17).

Evidently, vectors $\boldsymbol{d}_1 = \boldsymbol{a} + \boldsymbol{b}$ and $\boldsymbol{d}_2 = \boldsymbol{a} - \boldsymbol{b}$ represent the diagonals. If the diagonals are perpendicular their scalar product has to be zero, i.e.

$$\begin{aligned}\boldsymbol{d}_1 \cdot \boldsymbol{d}_2 &= (\boldsymbol{a}+\boldsymbol{b})\cdot(\boldsymbol{a}-\boldsymbol{b}) = \boldsymbol{a}\cdot\boldsymbol{a} + \boldsymbol{b}\cdot\boldsymbol{a} - \boldsymbol{a}\cdot\boldsymbol{b} - \boldsymbol{b}\cdot\boldsymbol{b} \\ &= a^2 - b^2 = 0.\end{aligned}$$

Thus $a = b$, implying that the parallelogram is a rhombus. ■

Example 4.3.16

Show that the sum of the squares of the diagonals of any parallelogram $ABCD$ is equal to the sum of the squares of the sides.

Solution

Consider a parallelogram $ABCD$ as in Figure 4.18:

Let $\overrightarrow{AB} = \boldsymbol{a}$ and $\overrightarrow{BC} = \boldsymbol{b}$. Then $\overrightarrow{DA} = -\boldsymbol{b}$, $\boldsymbol{d}_1 = \overrightarrow{AC} = \boldsymbol{a} + \boldsymbol{b}$ and $\boldsymbol{d}_2 = \overrightarrow{DB} = \boldsymbol{a} - \boldsymbol{b}$. So we have

$$d_1^2 = (\boldsymbol{a}+\boldsymbol{b})\cdot(\boldsymbol{a}+\boldsymbol{b}) = a^2 + b^2 + 2\boldsymbol{a}\cdot\boldsymbol{b}$$

and

$$d_2^2 = (\boldsymbol{a}-\boldsymbol{b})\cdot(\boldsymbol{a}-\boldsymbol{b}) = a^2 + b^2 - 2\boldsymbol{a}\cdot\boldsymbol{b}.$$

Hence,

$$d_1^2 + d_2^2 = 2a^2 + 2b^2.$$ ■

As mentioned before, because of their particular nature, vectors can be multiplied in yet another way. We address this next.

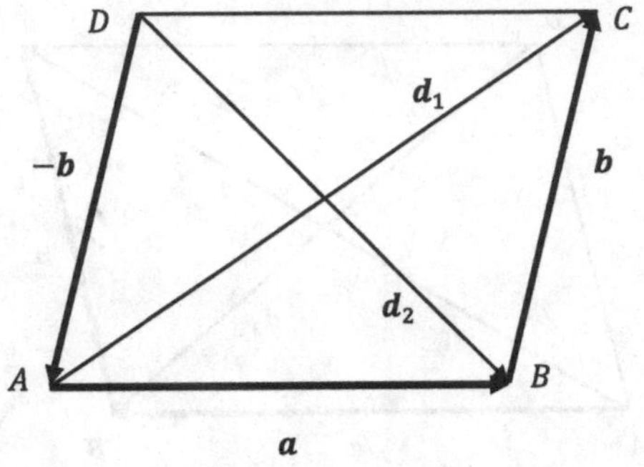

FIGURE 4.18

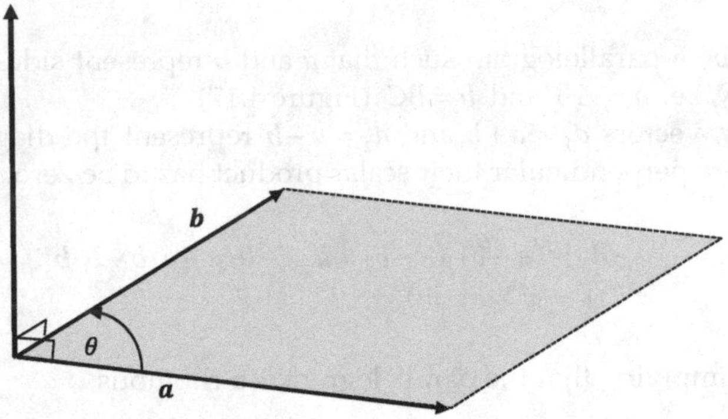

FIGURE 4.19

4.4 Cross (Vector) Product of Vectors

Definition 4.4.1 (Cross product of vectors)
Let $\boldsymbol{a}, \boldsymbol{b} \in \mathbf{R}^3$ be any two vectors. Then there exists a unique function

$$v : \mathbf{R}^3 \times \mathbf{R}^3 \to \mathbf{R}^3,$$

defined by

$$v(\boldsymbol{a}, \boldsymbol{b}) = \boldsymbol{a} \times \boldsymbol{b} = \boldsymbol{c},$$

called the ***cross product***. The cross product has the following properties:

(i) $\boldsymbol{a} \times \boldsymbol{b}$ is the vector $\boldsymbol{c}$ perpendicular to both $\boldsymbol{a}$ and $\boldsymbol{b}$, i.e. $\boldsymbol{c} \perp \boldsymbol{a}$, $\boldsymbol{c} \perp \boldsymbol{b}$ (Figure 4.19);
(ii) Vectors $\boldsymbol{a}, \boldsymbol{b}$ and $\boldsymbol{c}$ form a so-called *right-handed system*; and
(iii) The magnitude of $\boldsymbol{c}$ is equal to $ab \sin\theta$, where θ is the angle between vectors $\boldsymbol{a}$ and $\boldsymbol{b}$, i.e.

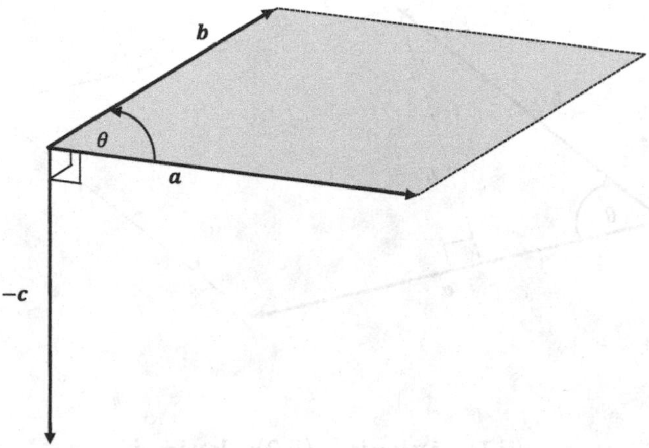

FIGURE 4.20

$$|c| = |\boldsymbol{a}\times\boldsymbol{b}| = ab\sin\theta, \quad 0\le\theta\le\pi.$$

Proposition 4.4.1
For every $\boldsymbol{a},\boldsymbol{b}\in\mathbf{R}^3$ the cross product is anticommutative (Figure 4.20), i.e.

$$\boldsymbol{a}\times\boldsymbol{b} = -(\boldsymbol{b}\times\boldsymbol{a})$$

Notice that the magnitude of **c**, i.e., $ab\sin\theta$, is equal to the area $\mathcal{A}$ of the parallelogram formed by vectors ***a*** and ***b***. Obviously, if ***a*** and ***b*** are collinear, then $\boldsymbol{c} = 0$.

Example 4.4.1
As a simple example let's justify (iii) from Definition 4.4.1.
The area $\mathcal{A}$ of any parallelogram (Figure 4.21) is $base\times height = a\cdot h$, so we have

$$\mathcal{A} = base\times height = a\cdot h = a\cdot b\sin\theta = |\boldsymbol{a}\times\boldsymbol{b}| = |c|.$$ ∎

Exercise 4.4.1
Show that (i) from Definition 4.4.1 above is indeed true, i.e., that for any two vectors $\boldsymbol{a},\boldsymbol{b}\in\mathbf{R}^3$, $\boldsymbol{a}\times\boldsymbol{b}$ is perpendicular to ***a*** and ***b***.

Exercise 4.4.2
Following the definition of cross vector product show that:

$$\text{(i.1)}\quad \mathbf{i}\times\mathbf{j}=\mathbf{k}, \qquad \text{(i.2)}\quad \mathbf{j}\times\mathbf{i}=-\mathbf{k},$$

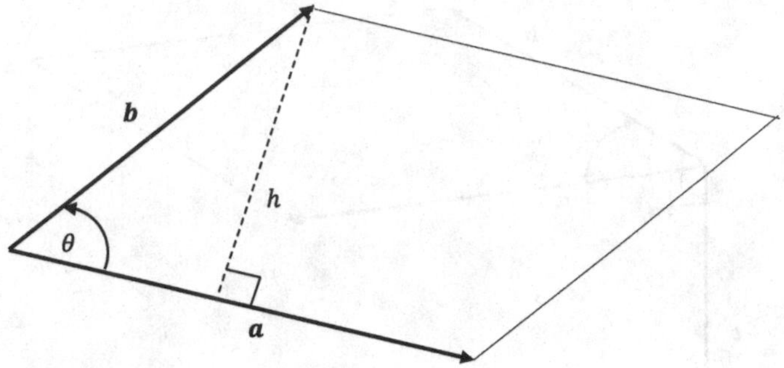

FIGURE 4.21

$$\text{(ii.1)} \quad \mathbf{j}\times\mathbf{k}=\mathbf{i}, \qquad \text{(ii.2)} \quad \mathbf{k}\times\mathbf{j}=-\mathbf{i},$$

$$\text{(iii.1)} \quad \mathbf{k}\times\mathbf{i}=\mathbf{j}, \qquad \text{(iii.2)} \quad \mathbf{i}\times\mathbf{k}=-\mathbf{j},$$

$$\text{(iv)} \quad \mathbf{i}\times\mathbf{i}=\mathbf{j}\times\mathbf{j}=\mathbf{k}\times\mathbf{k}=\mathbf{0}.$$

The proofs of the next two propositions follow immediately from the definition of the cross product.

Proposition 4.4.2
For any vector $\boldsymbol{a}$,

$$\mathbf{a}\times\mathbf{a}=\mathbf{0}.$$

Proof
Left to the reader. ■

Proposition 4.4.3
Two non-zero vectors $\boldsymbol{a},\boldsymbol{b}\in\mathbf{R}^3$ are collinear iff $\mathbf{a}\times\mathbf{b}=\mathbf{0}$.

Proof
Left to the reader.

Exercise 4.4.3
Show that $\mathbf{a}\times\mathbf{b}=\mathbf{0}$ iff there are scalars α and β, not both equal to zero, such that $\alpha\boldsymbol{a}=\beta\boldsymbol{b}$.

Exercise 4.4.4
Let $\boldsymbol{a}=(1,2,3)$ and $\boldsymbol{b}=(2,3,5)$. Find $\boldsymbol{a}\times\boldsymbol{b}$.

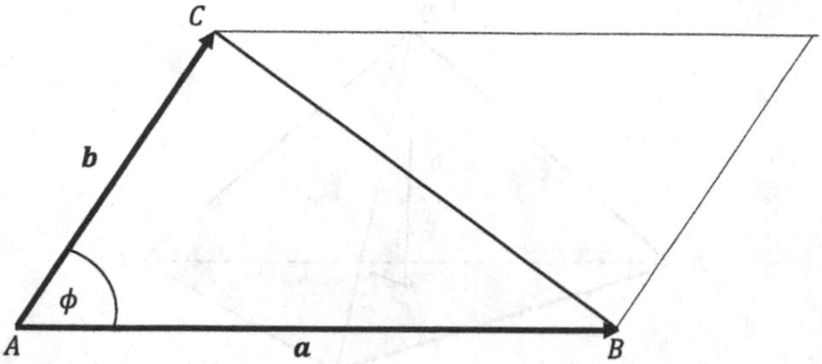

FIGURE 4.22

Exercise 4.4.5
Let $\boldsymbol{a}, \boldsymbol{b}, \boldsymbol{c} \in \mathbf{R}^3$ be any three vectors. Show that $\boldsymbol{a} \times \boldsymbol{b}$ can equal $\boldsymbol{a} \times \boldsymbol{c}$ without $\boldsymbol{b}$ being equal to $\boldsymbol{c}$.

Exercise 4.4.6
Let $\boldsymbol{a}, \boldsymbol{b}, \boldsymbol{c} \in \mathbf{R}^3$ with $\mathbf{a} \neq \mathbf{0}$. Show that if $\boldsymbol{a} \cdot \boldsymbol{b} = \boldsymbol{a} \cdot \boldsymbol{c}$ and $\boldsymbol{a} \times \boldsymbol{b} = \boldsymbol{a} \times \boldsymbol{c}$, then $\boldsymbol{b} = \boldsymbol{c}$.

Example 4.4.2
Let $A, B,$ and C be the vertices of a triangle ΔABC, so that $\overrightarrow{AB} = \boldsymbol{a}$ and $\overrightarrow{AC} = \boldsymbol{b}$ (Figure 4.22).

Let's derive the well-known formula for the area A of the triangle ΔABC.

Observe that the area A of the triangle is equal to one half of the area of the parallelogram $ABCD$. Since the area of the parallelogram is $|\overrightarrow{AB} \times \overrightarrow{AC}| = |\boldsymbol{a} \times \boldsymbol{b}|$ it follows that

$$\mathcal{A} = \frac{1}{2}|\boldsymbol{a} \times \boldsymbol{b}| = \frac{1}{2} ab \sin \phi. \qquad \blacksquare$$

Exercise 4.4.7
Find the area of the triangle ΔABC if $\overrightarrow{AB} = 2\boldsymbol{i} - 3\boldsymbol{k}$ and $\overrightarrow{AC} = 4\boldsymbol{i} + \boldsymbol{j} + 4\boldsymbol{k}$.

Exercise 4.4.8
Consider a tetrahedron with vertices $A(1,0,2), B(3,-1,4), C(1,5,2)$ and $D(4,4,4)$, (Figure 4.23).
Show that:

(i) the height h of the tetrahedron is $\dfrac{\sqrt{2}}{2}$.

(ii) the volume V of the tetrahedron is $\dfrac{5}{3}$. $\left(\textbf{Hint}: \mathrm{V} = \frac{1}{3}(\textit{area of the base}) \cdot (\textit{height})\right)$

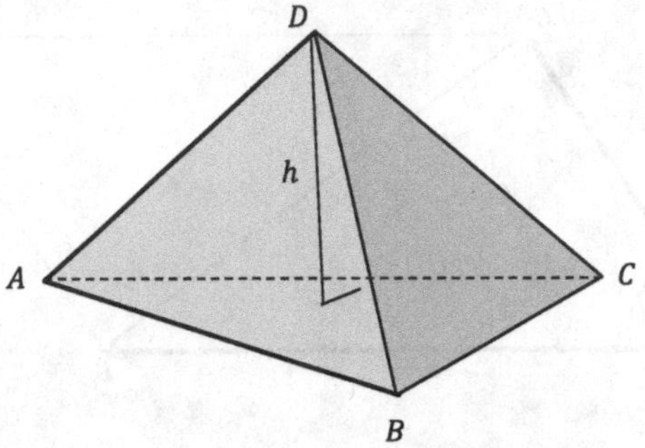

FIGURE 4.23

Proposition 4.4.4
Let $\boldsymbol{a} = \alpha_x \boldsymbol{i} + \alpha_y \boldsymbol{j} + \alpha_z \boldsymbol{k}$ and $\boldsymbol{b} = \beta_x \boldsymbol{i} + \beta_y \boldsymbol{j} + \beta_z \boldsymbol{k}$. Then

$$\boldsymbol{a} \times \boldsymbol{b} = \left(\alpha_y \beta_z - \alpha_z \beta_y\right)\boldsymbol{i} - \left(\alpha_x \beta_z - \alpha_z \beta_x\right)\boldsymbol{j} + \left(\alpha_x \beta_y - \alpha_y \beta_x\right)\boldsymbol{k}.$$

Proof

$$\begin{aligned}
\boldsymbol{a} \times \boldsymbol{b} &= \left(\alpha_x \boldsymbol{i} + \alpha_y \boldsymbol{j} + \alpha_z \boldsymbol{k}\right) \times \left(\beta_x \boldsymbol{i} + \beta_y \boldsymbol{j} + \beta_z \boldsymbol{k}\right) \\
&= \alpha_x \beta_x (\boldsymbol{i} \times \boldsymbol{i}) + \alpha_x \beta_y (\boldsymbol{i} \times \boldsymbol{j}) + \alpha_x \beta_z (\boldsymbol{i} \times \boldsymbol{k}) \\
&+ \alpha_y \beta_x (\boldsymbol{j} \times \boldsymbol{i}) + \alpha_y \beta_x (\boldsymbol{j} \times \boldsymbol{j}) + \alpha_y \beta_z (\boldsymbol{j} \times \boldsymbol{k}) \\
&+ \alpha_z \beta_x (\boldsymbol{k} \times \boldsymbol{i}) + \alpha_z \beta_{xy} (\boldsymbol{k} \times \boldsymbol{j}) + \alpha_z \beta_z (\boldsymbol{k} \times \boldsymbol{k}) \\
&= \alpha_x \beta_y \boldsymbol{k} + \alpha_x \beta_z (-\boldsymbol{j}) + \alpha_y \beta_x (-\boldsymbol{k}) + + \alpha_y \beta_z \boldsymbol{i} + + \alpha_z \beta_x \boldsymbol{j} + \alpha_z \beta_y (-\boldsymbol{i}) \\
&= (\alpha_y \beta_z - \alpha_z \beta_y)\boldsymbol{i} - \left(\alpha_x \beta_z - \alpha_z \beta_x\right)\boldsymbol{j} + \left(\alpha_x \beta_y - \alpha_y \beta_x\right)\boldsymbol{k}.
\end{aligned}$$

■

Example 4.4.3
Find $\boldsymbol{a} \times \boldsymbol{b}$ if $\boldsymbol{a} = 2\boldsymbol{i} - \boldsymbol{j} + \boldsymbol{k}$ and $\boldsymbol{b} = \boldsymbol{i} - 2\boldsymbol{k}$.

Solution

$$\begin{aligned}
\boldsymbol{a} \times \boldsymbol{b} &= (2\boldsymbol{i} - \boldsymbol{j} + \boldsymbol{k}) \times (\boldsymbol{i} - 2\boldsymbol{k}) \\
&= 2\boldsymbol{i} \times (\boldsymbol{i} - 2\boldsymbol{k}) - \boldsymbol{j} \times (\boldsymbol{i} - 2\boldsymbol{k}) + \boldsymbol{k} \times (\boldsymbol{i} - 2\boldsymbol{k}) \\
&= 2(\boldsymbol{i} \times \boldsymbol{i}) - 4(\boldsymbol{i} \times \boldsymbol{k}) - (\boldsymbol{j} \times \boldsymbol{i}) + 2(\boldsymbol{j} \times \boldsymbol{k}) + (\boldsymbol{k} \times \boldsymbol{i}) - 2(\boldsymbol{k} \times \boldsymbol{k}) \\
&= 4\boldsymbol{j} + \boldsymbol{k} + 2\boldsymbol{i} + \boldsymbol{j} \\
&= 2\boldsymbol{i} + 5\boldsymbol{j} + \boldsymbol{k}
\end{aligned}$$

Since the components of the cross product are, according to Proposition 4.4.4:

$$(\boldsymbol{a} \times \boldsymbol{b})_x = \left(\alpha_y \beta_z - \alpha_z \beta_y\right),$$

$$(a \times b)_y = (\alpha_z \beta_x - \alpha_x \beta_z),$$

$$(a \times b)_z = (\alpha_x \beta_y - \alpha_y \beta_x),$$

a reader familiar with determinants[3] will immediately recognize that there is a simple way of calculating the cross product of two vectors a and b, which is by evaluating the determinant:

$$a \times b = \begin{vmatrix} i & j & k \\ \alpha_x & \alpha_y & \alpha_z \\ \beta_x & \beta_y & \beta_z \end{vmatrix} = (\alpha_y \beta_z - \alpha_z \beta_y) i - (\alpha_z \beta_x - \alpha_x \beta_z) j + (\alpha_x \beta_y - \alpha_y \beta_x) k$$

Example 4.4.4

Find $a \times b$ by the "determinant method" if $a = 2i + 4j - 3k$ and $b = i + 3j + 7k$.

Solution

$$a \times b = \begin{vmatrix} i & j & k \\ 2 & 4 & -3 \\ 1 & 3 & 7 \end{vmatrix} = (4 \cdot 7 - (-3) \cdot 3) i - (2 \cdot 7 - (-3) \cdot 1) j + (2 \cdot 3 - 4 \cdot 1) k = 37i - 17j + 2k.$$

■

Exercise 4.4.9

Let $a = 3i + 2j + k$ and $b = i - 3j + 4k$. Show that

$$a \times b = \begin{vmatrix} i & j & k \\ 3 & 2 & 1 \\ 1 & -3 & 4 \end{vmatrix} = \ldots = 11i - 11j - 11k,$$

and

$$b \times a = \begin{vmatrix} i & j & k \\ 1 & -3 & 4 \\ 3 & 2 & 1 \end{vmatrix} = \ldots = -11i + 11j + 11k = -(b \times a).$$

Example 4.4.5

Here is the vector product of the unit vectors in (i.1) and (ii.2) from Exercise 4.4.2 evaluated by the "determinant method":

Since $\boldsymbol{i} = (1,0,0)$, $\boldsymbol{j} = (0,1,0)$ and $\boldsymbol{k} = (0,0,1)$, we have

$$\boldsymbol{i}\times\boldsymbol{j} = \begin{vmatrix} \boldsymbol{i} & \boldsymbol{j} & \boldsymbol{k} \\ 1 & 0 & 0 \\ 0 & 1 & 0 \end{vmatrix} = \boldsymbol{k}; \qquad \boldsymbol{k}\times\boldsymbol{j} = \begin{vmatrix} \boldsymbol{i} & \boldsymbol{j} & \boldsymbol{k} \\ 0 & 0 & 1 \\ 0 & 1 & 0 \end{vmatrix} = -\boldsymbol{j}.$$

■

Exercise 4.4.10
Verify that the results from the previous three examples can be obtained by calculating the cross product using the method described in Proposition 4.4.4.

Example 4.4.6
Let's look again at statement (i) from Definition 4.4.1.

Take two non-collinear vectors $\boldsymbol{a}, \boldsymbol{b}$ in the XY-plane, and let vector $\boldsymbol{a}$ be directed along the X-axis (Figure 4.24).

In other words,

$$\boldsymbol{a} = a_x\boldsymbol{i} = a\boldsymbol{i},$$

and let

$$\boldsymbol{b} = b_x\boldsymbol{i} + b_y\boldsymbol{j} = b\cos\theta\boldsymbol{i} + b\sin\theta\boldsymbol{j}.$$

Then

$$\boldsymbol{c} = \boldsymbol{a}\times\boldsymbol{b} = \begin{vmatrix} \boldsymbol{i} & \boldsymbol{j} & \boldsymbol{k} \\ a_x & 0 & 0 \\ b_x & b_y & 0 \end{vmatrix} = \begin{vmatrix} \boldsymbol{i} & \boldsymbol{j} & \boldsymbol{k} \\ a & 0 & 0 \\ b\cos\theta & b\sin\theta & 0 \end{vmatrix}$$
$$= ab\sin\theta\boldsymbol{k}.$$

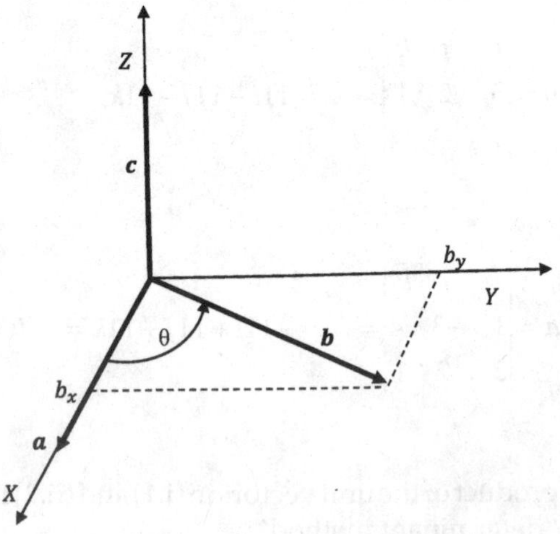

FIGURE 4.24

So, as expected, c is directed along the Z-axis, and therefore it is perpendicular to both a and b. ■

Proposition 4.4.5
The cross product has the following properties:

(i) $(\lambda a)\times b = a\times(\lambda b) = \lambda(a\times b), \forall \lambda \in \mathbf{R}, a, b \in \mathbf{R}^3$;
(ii) $a\times(b+c) = (a\times b)+(a\times c), \forall a, b, c \in \mathbf{R}^3$.

Proof
(i) If $\lambda = 0$ the claim is obvious.
If $\lambda > 0$, then a and λa have the same orientation, i.e. $\sphericalangle(\lambda a, b) = \sphericalangle(a, b) = \phi$. So

$$
\begin{aligned}
|(\lambda a)\times b| &= |\lambda a||b|\sin(\lambda a, b) \\
&= \lambda a\cdot b\cdot \sin(\lambda a, b) = \lambda ab\sin(a, b) \\
&= \lambda|a\times b| = |\lambda(a\times b)|.
\end{aligned}
$$

If $\lambda < 0$, then $\sphericalangle(\lambda a, b) = \pi - \sphericalangle(a, b)$.
So

$$
\begin{aligned}
|(\lambda a)\times b| &= |\lambda a||b|\sin(\lambda a, b) \\
&= |\lambda a||b|\sin[\pi - \sphericalangle(a, b)] \\
&= -\lambda a\cdot b\cdot \sin(a, b) \\
&= -\lambda|a\times b| \\
&= |\lambda||a\times b| = |\lambda(a\times b)|.
\end{aligned}
$$

(ii) In our proof we will assume that $a \neq b \neq c \neq 0$; for if any of the vectors a, b and c was equal to 0 the claim would be trivial. So, let them be as in Figure 4.25.

Now, notice that

$$|\overrightarrow{OB'}| = |b|\cos\left(\frac{\pi}{2} - \phi\right) = b\sin\phi.$$

In other words, $|\overrightarrow{OB'}|$ is a projection of vector b to the plane Π. Let's rotate $\overrightarrow{OB'}$ counterclockwise by 90° to get $\overrightarrow{OC}$, which aligns with vector $a\times b$, i.e. $|\overrightarrow{OC}| = |\overrightarrow{OB'}|$. Observe, also, that $\overrightarrow{OC} \perp \overrightarrow{OB'}$. Multiplying a by $b\sin\phi$ yields $c = a\times b$.

Next, let $b + c = d$, so we are now dealing with vectors b, c and d. Finally, we consider the products $a\times b$, $a\times c$, and $a\times d$, and we get

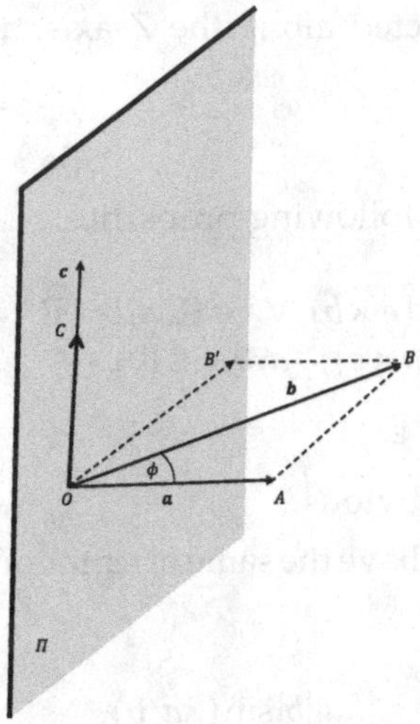

FIGURE 4.25

$$a \times b + a \times c = a \times d$$
$$= a \times (b + c).$$

■

Exercise 4.4.11
Prove that

$$(a \times \lambda b) = \lambda(a \times b)$$

(cf. Proposition 4.4.5).

Exercise 4.4.12
Prove that

$$(\alpha a + \beta b) \times c = \alpha(a \times c) + \beta(b \times c)$$

(cf. Proposition 4.4.5).

Exercise 4.4.13
Show that for any two vectors $a, b \in \mathbf{R}^3$, and $\alpha \in \mathbf{R}$, if $b = \alpha a + c$, then $a \times b = a \times c$.

Example 4.4.7
Prove that for any $a, b \in \mathbf{R}^3$

$$(a - b) \times (a + b) = 2(a \times b).$$

Solution

$$\begin{aligned}(a-b)\times(a+b) &= (a-b)\times a+(a-b)\times b\\ &= a\times a-b\times a+a\times b-b\times b\\ &= a\times b+a\times b\\ &= 2(a\times b).\end{aligned}$$

■

Exercise 4.4.14
Let $a = 2i-3j-k$ and $b = i+4j-2k$. Show that

$$(a+b)\times(a-b) = -20i-6j-22k.$$

Example 4.4.8
Consider a parallelogram Π_1 whose sides are the vectors a and b (Figure 4.26). Show that the area $\mathcal{A}(\Pi_2)$ of the parallelogram Π_2 whose sides are the diagonals of Π_1 is twice the area $\mathcal{A}(\Pi_1)$, i.e. $\mathcal{A}(\Pi_2) = 2\mathcal{A}(\Pi_1)$.

Solution
Note that the area of parallelogram Π_1 is $\mathcal{A}(\Pi_1) = |a\times b|$, while the area $\mathcal{A}(\Pi_2) = |(a+b)\times(a-b)|$. Thus,

$$\begin{aligned}|(a+b)\times(a-b)| &= |(a+b)\times a-(a+b)\times b|\\ &= |(a\times a)+(b\times a)-(a\times b)-(b\times b)|\\ &= |(b\times a)+(b\times a)|\\ &= 2|(b\times a)|.\end{aligned}$$

■

Example 4.4.9
Let a and b be coplanar with $\measuredangle(a,b) = \theta$. Furthermore, let b' be the projection vector of the vector b on the line *perpendicular* to a, i.e. $b' \perp a$ (Figure 4.27). Show that $a\times b = a\times b'$.

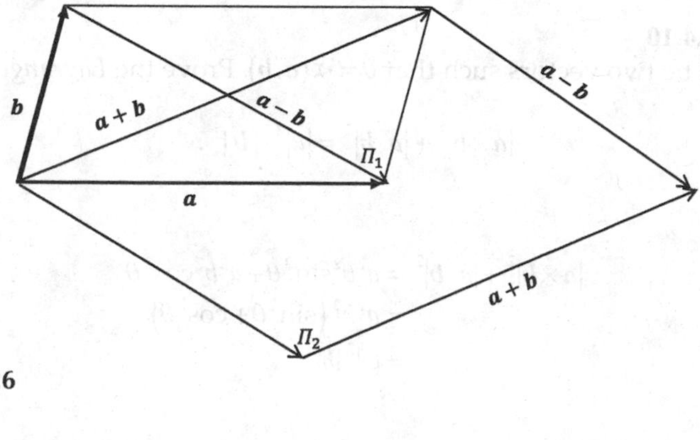

FIGURE 4.26

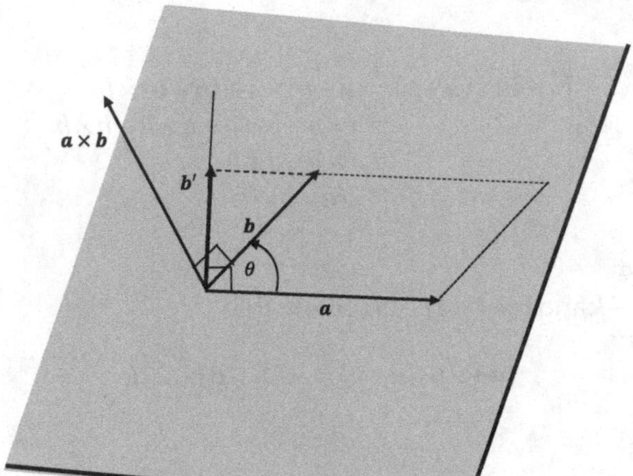

FIGURE 4.27

Solution
Since

$$|\boldsymbol{a}\times\boldsymbol{b}| = |a\cdot b\sin(\boldsymbol{a},\boldsymbol{b})| = |a\cdot b\sin\theta|$$

and

$$|b|\sin\theta = |b|\,|\cos\left(\frac{\pi}{2}-\theta\right) = |b'|,$$

it follows that

$$|\boldsymbol{a}\times\boldsymbol{b}'| = |a\cdot b'\sin(\boldsymbol{a},\boldsymbol{b}')| = |a\cdot b'|.$$

Thus

$$\boldsymbol{a}\times\boldsymbol{b} = \boldsymbol{a}\times\boldsymbol{b}'.$$ ■

Example 4.4.10
Let $\boldsymbol{a}$ and $\boldsymbol{b}$ be two vectors such that $\theta = \sphericalangle(\boldsymbol{a},\boldsymbol{b})$. Prove the *Lagrange identity*:

$$|\boldsymbol{a}\times\boldsymbol{b}|^2 + |\boldsymbol{a}\cdot\boldsymbol{b}|^2 = |\boldsymbol{a}|^2\cdot|\boldsymbol{b}|^2.$$

Solution

$$\begin{aligned}|\boldsymbol{a}\times\boldsymbol{b}|^2 + |\boldsymbol{a}\cdot\boldsymbol{b}|^2 &= a^2b^2\sin^2\theta + a^2b^2\cos^2\theta\\ &= a^2b^2\left(\sin^2\theta+\cos^2\theta\right)\\ &= |a|^2\,|b|^2.\end{aligned}$$ ■

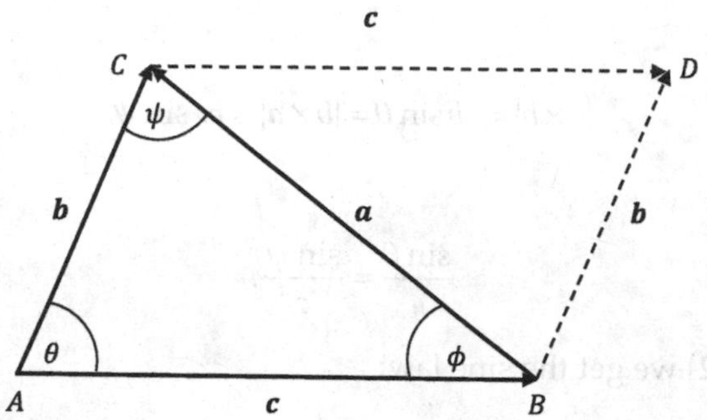

FIGURE 4.28

Example 4.4.11

Let's derive the sine law. Consider triangle ΔABC in Figure 4.28.

Notice that $\boldsymbol{a} = \boldsymbol{b} - \boldsymbol{c}$, so we can write

$$0 = \boldsymbol{a} \times \boldsymbol{a} = (\boldsymbol{b} - \boldsymbol{c}) \times \boldsymbol{a} = \boldsymbol{b} \times \boldsymbol{a} - \boldsymbol{c} \times \boldsymbol{a}.$$

Thus,

$$\boldsymbol{b} \times \boldsymbol{a} = \boldsymbol{c} \times \boldsymbol{a}$$

and

$$|\boldsymbol{b} \times \boldsymbol{a}| = |\boldsymbol{c} \times \boldsymbol{a}|.$$

But,

$$|\boldsymbol{b} \times \boldsymbol{a}| = ba \sin \psi = |\boldsymbol{c} \times \boldsymbol{a}| = ca \sin \phi,$$

and, since $a \neq 0$,

$$\frac{\sin \psi}{c} = \frac{\sin \phi}{b}. \tag{4.4}$$

Similarly,

$$0 = \boldsymbol{b} \times \boldsymbol{b} = \boldsymbol{b} \times (\boldsymbol{c} + \boldsymbol{a}) = \boldsymbol{b} \times \boldsymbol{c} + \boldsymbol{b} \times \boldsymbol{a}$$

Thus,

$$\boldsymbol{c} \times \boldsymbol{b} = \boldsymbol{b} \times \boldsymbol{a}$$

and

$$|\boldsymbol{c} \times \boldsymbol{b}| = |\boldsymbol{b} \times \boldsymbol{a}|.$$

So,

$$|c \times b| = cb\sin\theta = |b \times a| = ba\sin\psi$$

and

$$\frac{\sin\theta}{a} = \frac{\sin\psi}{c}. \tag{4.5}$$

From (1) and (2) we get the sine law:

$$\frac{\sin\theta}{a} = \frac{\sin\phi}{b} = \frac{\sin\psi}{c}$$

■

Example 4.4.12
Let's prove another well-known trigonometric identity (cf. Example 4.3.13), namely

$$\sin(\theta - \phi) = \sin\theta\cos\phi - \cos\theta\sin\phi$$

Solution
Let $\boldsymbol{a}$ and $\boldsymbol{b}$ be two vectors in the first quadrant making angles θ and ϕ with the X axis and, without loss of generality, let $a = b = 1$ (Figure 4.29):

Then

$$\boldsymbol{a} = \cos\theta\boldsymbol{i} + \sin\theta\boldsymbol{j}$$

and

$$\boldsymbol{b} = \cos\phi\boldsymbol{i} + \sin\phi\boldsymbol{j}.$$

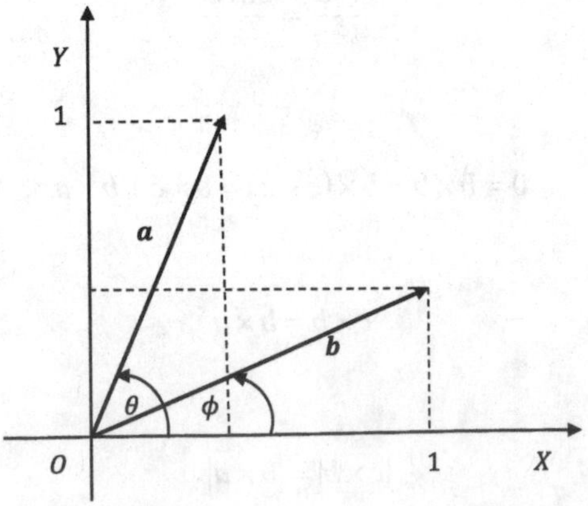

FIGURE 4.29

So, the cross product is

$$\begin{aligned} \boldsymbol{a}\times\boldsymbol{b} &= (\cos\phi\boldsymbol{i}+\sin\phi\boldsymbol{j})\times(\cos\theta\boldsymbol{i}+\sin\theta\boldsymbol{j}) \\ &= \begin{vmatrix} \boldsymbol{i} & \boldsymbol{j} & \boldsymbol{k} \\ \cos\phi & \sin\phi & 0 \\ \cos\theta & \sin\theta & 0 \end{vmatrix} \\ &= (\sin\theta\cos\phi-\cos\theta\sin\phi)\boldsymbol{k}. \end{aligned}$$

On the other hand,

$$\begin{aligned} |\boldsymbol{a}\times\boldsymbol{b}| &= 1\cdot 1\sin(\theta-\phi) \\ &= \sin(\theta-\phi) \end{aligned}$$

Hence,

$$\sin(\theta-\phi)=\sin\theta\cos\phi-\cos\theta\sin\phi. \qquad \blacksquare$$

Example 4.4.13
Consider a rigid body rotating with a constant angular velocity *w* about the *Z*-axis (Figure 4.30), and find its linear velocity ***v***.

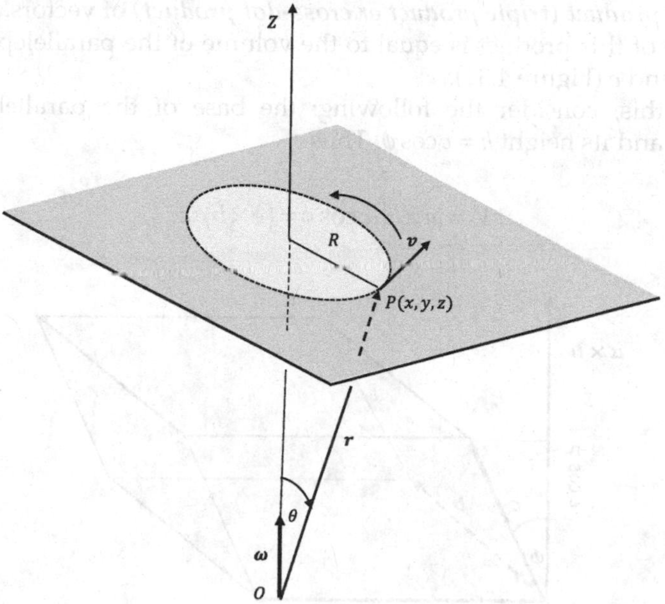

FIGURE 4.30

Solution
Let $\boldsymbol{r}$ be the position vector of the point $P(x,y,z)$ on the path of circular rotation. From the figure we see that the vector $\boldsymbol{v}$ lies in the plane of rotation, that is, it is perpendicular to the plane of $\boldsymbol{w}$ and $\boldsymbol{r}$ (i.e. it is perpendicular to both $\boldsymbol{w}$ and $\boldsymbol{r}$). Thus, since $R = r\sin\theta$ is the radius of rotation, the magnitude of $\boldsymbol{v}$ is

$$v = \omega R = \omega \cdot r\sin\theta = |\boldsymbol{w}\times\boldsymbol{r}|.$$

So, $\boldsymbol{v} = \boldsymbol{w}\times\boldsymbol{r}$. ■

4.5 Mixed Product of Vectors

Definition 4.5.1 (Mixed product of vectors)
We say that the map

$$m : \mathbf{R}^3 \times \mathbf{R}^3 \times \mathbf{R}^3 \to \mathbf{R},$$

defined as

$$m(\boldsymbol{a},\boldsymbol{b},\boldsymbol{c}) = (\boldsymbol{a}\times\boldsymbol{b})\cdot\boldsymbol{c},$$

is a ***mixed product** (**triple product** or **cross-dot product**)* of vectors. The absolute value of this product is equal to the volume of the parallelepiped with sides $\boldsymbol{a},\boldsymbol{b}$ and $\boldsymbol{c}$ (Figure 4.31).

To see this, consider the following: the base of the parallelepiped is $B = |\boldsymbol{a}\times\boldsymbol{b}|$, and its height $h = c\cos\phi$. Thus

$$V = |\boldsymbol{a}\times\boldsymbol{b}|\cdot c\cos\phi = (\boldsymbol{a}\times\boldsymbol{b})\cdot\boldsymbol{c}.$$

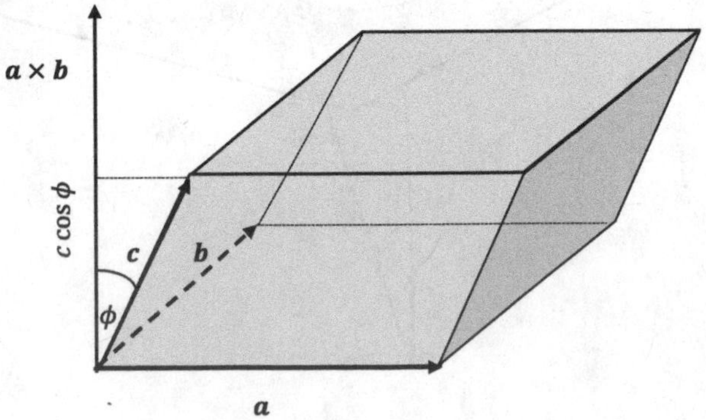

FIGURE 4.31

Example 4.5.1
Find the volume of the parallelepiped formed by

$$a = 3i + 2k, b = i + 2j + k \quad \text{and} \quad c = -j + 4k.$$

Solution

$$a \cdot (b \times c) = \begin{vmatrix} 3 & 0 & 2 \\ 1 & 2 & 1 \\ 0 & -1 & 4 \end{vmatrix} = 3(8+1) + 2(-1) = 25.$$

So the volume is 25 cubic units. ■

Exercise 4.5.1
Prove that the volume of the parallelepiped formed by $a = 2i - 3j, b = i + j - k$ and $c = 3i - k$ is 4 cubic units.

Exercise 4.5.2
Prove that $(a \times b) \cdot c = c \cdot (a \times b)$.

Proposition 4.5.1
For any $a = \alpha_x i + \alpha_y j + \alpha_z k, b = \beta_x i + \beta_y j + \beta_z k$ and $c = \gamma_x i + \gamma_y j + \gamma_z k$,

$$(a \times b) \cdot c = (b \times c) \cdot a = (c \times a) \cdot b.$$

Proof
Let $a \times b = d$. Then

$$(a \times b) \cdot c = d \cdot c = d_x \gamma_x + d_y \gamma_y + d_z \gamma_z. \tag{4.6}$$

But

$$a \times b = (\alpha_y \beta_z - \alpha_z \beta_y) i + (\alpha_z \beta_x - \alpha_x \beta_z) j + (\alpha_x \beta_y - \alpha_y \beta_x) k. \tag{4.7}$$

So, we have

$$(a \times b) \cdot c = \gamma_x (\alpha_y \beta_z - \alpha_z \beta_y) + \gamma_y (\alpha_z \beta_x - \alpha_x \beta_z) + \gamma_z (\alpha_x \beta_y - \alpha_y \beta_x). \tag{4.8}$$

In Equation (4.3) we recognize the determinant

$$D = \begin{vmatrix} \alpha_x & \alpha_y & \alpha_z \\ \beta_x & \beta_y & \beta_z \\ \gamma_x & \gamma_y & \gamma_z \end{vmatrix}$$

■

evaluated on the third row. From elementary theorems on determinants (see Appendix C), it can be shown that interchanging two rows produces a change of sign of the determinant, so that an even number of such interchanges leaves the value of the determinant unchanged. Hence the cyclic permutation of the vectors does not change the value of the dot-cross product, i.e.

$$\boldsymbol{a}\cdot(\boldsymbol{b}\times\boldsymbol{c})=\boldsymbol{b}\cdot(\boldsymbol{c}\times\boldsymbol{a})=\boldsymbol{c}\cdot(\boldsymbol{a}\times\boldsymbol{b}), \quad (*)$$

which we often write as

$$(\boldsymbol{a},\boldsymbol{b},\boldsymbol{c})=\begin{vmatrix}\alpha_x & \alpha_y & \alpha_z\\ \beta_x & \beta_y & \beta_z\\ \gamma_x & \gamma_y & \gamma_z\end{vmatrix}.$$

Example 4.5.2
Show that for any three vectors $\boldsymbol{a},\boldsymbol{b},\boldsymbol{c}\in\mathbf{R}^3$

$$\boldsymbol{a}\cdot(\boldsymbol{b}\times\boldsymbol{c})=(\boldsymbol{a}\times\boldsymbol{b})\cdot\boldsymbol{c}.$$

Solution
Since the scalar product of two vectors is commutative it follows that

$$\boldsymbol{c}\cdot(\boldsymbol{a}\times\boldsymbol{b})=(\boldsymbol{a}\times\boldsymbol{b})\cdot\boldsymbol{c}.$$

On the other hand, from Proposition 4.5.1 it follows that

$$\boldsymbol{a}\cdot(\boldsymbol{b}\times\boldsymbol{c})=(\boldsymbol{a}\times\boldsymbol{b})\cdot\boldsymbol{c}. \qquad \blacksquare$$

Finally, from everything that has been said so far, it is fairly obvious that:

$$(\boldsymbol{a},\boldsymbol{b},\boldsymbol{c})=-(\boldsymbol{b},\boldsymbol{a},\boldsymbol{c})=-(\boldsymbol{c},\boldsymbol{b},\boldsymbol{a})=-(\boldsymbol{a},\boldsymbol{c},\boldsymbol{b}).$$

Corollary 4.5.1
(i) $(\alpha\boldsymbol{a},\beta\boldsymbol{b},\gamma\boldsymbol{c})=\alpha\beta\gamma(\boldsymbol{a},\boldsymbol{b},\boldsymbol{c})$;
(ii) $(\boldsymbol{a}_1+\boldsymbol{a}_2,\boldsymbol{b},\boldsymbol{c})=(\boldsymbol{a}_1,\boldsymbol{b},\boldsymbol{c})+(\boldsymbol{a}_2,\boldsymbol{b},\boldsymbol{c})$;
(iii) $(\boldsymbol{a},\boldsymbol{b}_1+\boldsymbol{b}_2,\boldsymbol{c})=(\boldsymbol{a},\boldsymbol{b}_1,\boldsymbol{c})+(\boldsymbol{a},\boldsymbol{b}_2,\boldsymbol{c})$; and
(iv) $(\boldsymbol{a},\boldsymbol{b},\boldsymbol{c}_1+\boldsymbol{c}_2)=(\boldsymbol{a},\boldsymbol{b},\boldsymbol{c}_1)+(\boldsymbol{a},\boldsymbol{b},\boldsymbol{c}_2)$.

Exercise 4.5.3
Let $\boldsymbol{a}=(4,-2,1)$, $\boldsymbol{b}=(1,-1,-3)$ and $\boldsymbol{c}=(2,1,2)$. Evaluate $\boldsymbol{a}\cdot(\boldsymbol{b}\times\boldsymbol{c})$.

Exercise 4.5.4
Prove that for any $\boldsymbol{a},\boldsymbol{b}\in\mathbf{R}^3$, $\boldsymbol{a}\cdot(\boldsymbol{a}\times\boldsymbol{b})=0$.

Exercise 4.5.5
Show that:

(i) $\boldsymbol{i}\cdot(\boldsymbol{j}\times\boldsymbol{k})=1$;
(ii) $\boldsymbol{j}\cdot(\boldsymbol{j}\times\boldsymbol{i})=0$.

Example 4.5.3
Let $\boldsymbol{x}$ be any vector, and let

$$\boldsymbol{y}=\boldsymbol{a}\times(\boldsymbol{b}+\boldsymbol{c})-\boldsymbol{a}\times\boldsymbol{b}-\boldsymbol{a}\times\boldsymbol{c}.$$

Show that $\boldsymbol{x}\cdot\boldsymbol{y}=0$.

Solution

$$\begin{aligned}\boldsymbol{x}\cdot\boldsymbol{y}&=\boldsymbol{x}\cdot\left[\boldsymbol{a}\times(\boldsymbol{b}+\boldsymbol{c})-\boldsymbol{a}\times\boldsymbol{b}-\boldsymbol{a}\times\boldsymbol{c}\right]\\&=\boldsymbol{x}\cdot\left[\boldsymbol{a}\times(\boldsymbol{b}+\boldsymbol{c})\right]-\boldsymbol{x}\cdot(\boldsymbol{a}\times\boldsymbol{b})-\boldsymbol{x}\cdot(\boldsymbol{a}\times\boldsymbol{c})\\&=(\boldsymbol{x}\times\boldsymbol{a})\cdot(\boldsymbol{b}+\boldsymbol{c})-(\boldsymbol{x}\times\boldsymbol{a})\cdot\boldsymbol{b}-(\boldsymbol{x}\times\boldsymbol{a})\cdot\boldsymbol{c}\\&=(\boldsymbol{x}\times\boldsymbol{a})\cdot\boldsymbol{b}+(\boldsymbol{x}\times\boldsymbol{a})\cdot\boldsymbol{c}-(\boldsymbol{x}\times\boldsymbol{a})\cdot\boldsymbol{b}-(\boldsymbol{x}\times\boldsymbol{a})\cdot\boldsymbol{c}\\&=0.\end{aligned}$$ ■

Proposition 4.5.2
Three vectors $\boldsymbol{a},\boldsymbol{b},\boldsymbol{c}$ are coplanar iff $\boldsymbol{a}\cdot(\boldsymbol{b}\times\boldsymbol{c})=0$.

Proof
If $\boldsymbol{a},\boldsymbol{b},\boldsymbol{c}$ are coplanar, the volume of parallelepiped formed by them is zero, and so $\boldsymbol{a}\cdot(\boldsymbol{b}\times\boldsymbol{c})=0$. On the other hand, if $\boldsymbol{a}\cdot(\boldsymbol{b}\times\boldsymbol{c})=0$ then the volume of the parallelepiped formed by $\boldsymbol{a},\boldsymbol{b}$ and $\boldsymbol{c}$ is zero, so the vectors must lie in a plane.

Corollary 4.5.2
Vectors $\boldsymbol{a}=\alpha_x\boldsymbol{i}+\alpha_y\boldsymbol{j}+\alpha_z\boldsymbol{k}$, $\boldsymbol{b}=\beta_x\boldsymbol{i}+\beta_y\boldsymbol{j}+\beta_z\boldsymbol{k}$ and $\boldsymbol{c}=\gamma_x\boldsymbol{i}+\gamma_y\boldsymbol{j}+\gamma_z\boldsymbol{k}$ are coplanar iff

$$\begin{vmatrix}\alpha_x & \alpha_y & \alpha_z\\ \beta_x & \beta_y & \beta_z\\ \gamma_x & \gamma_y & \gamma_z\end{vmatrix}=0.$$

Equivalently, we say that the vectors $\boldsymbol{a},\boldsymbol{b},\boldsymbol{c}\in\mathbf{R}^3$ are linearly dependent iff

$$\boldsymbol{a}\cdot(\boldsymbol{b}\times\boldsymbol{c})=0.$$

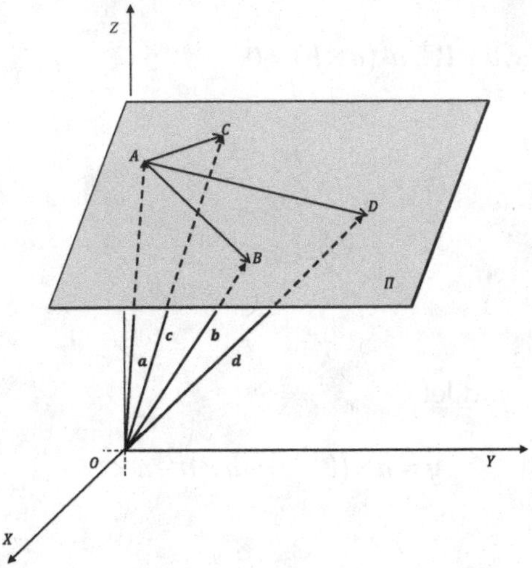

FIGURE 4.32

Example 4.5.4
Let's determine a condition under which four distinct points are coplanar. Consider four distinct points A, B, C and D lying in the plane Π, with corresponding position vectors $\boldsymbol{a}, \boldsymbol{b}, \boldsymbol{c}$ and $\boldsymbol{d}$ (Figure 4.32).

Since A, B, C and D are coplanar, $\overrightarrow{AC} \times \overrightarrow{AD}$ is a vector perpendicular to the plane Π, and therefore perpendicular to the vector $\overrightarrow{AB}$. Consequently

$$\overrightarrow{AB} \cdot \left(\overrightarrow{AC} \times \overrightarrow{AD}\right) = 0.$$

But since

$$\overrightarrow{AB} = \boldsymbol{b} - \boldsymbol{a},$$

$$\overrightarrow{AC} = \boldsymbol{c} - \boldsymbol{a},$$

and

$$\overrightarrow{AD} = \boldsymbol{d} - \boldsymbol{a},$$

the necessary condition for the coplanarity of four points is

$$(\boldsymbol{b} - \boldsymbol{a}) \cdot \left[(\boldsymbol{c} - \boldsymbol{a}) \times (\boldsymbol{d} - \boldsymbol{a})\right] = 0.$$

■

Example 4.5.5
Determine whether the points $A(-1,2,2), B(3,3,4), C(2,-2,10)$ and $D(0,2,2)$ are coplanar.

Solution
Let $\boldsymbol{a}=\overrightarrow{AB}=(4,1,2), \boldsymbol{b}=\overrightarrow{AC}=(3,4,8),\ \boldsymbol{c}=\overrightarrow{AD}=(1,0,0)$, and we have

$$\boldsymbol{a}\cdot(\boldsymbol{b}\times\boldsymbol{c})=\begin{vmatrix}4 & 1 & 2\\ 3 & 4 & 8\\ 1 & 0 & 0\end{vmatrix}=\begin{vmatrix}1 & 2\\ 4 & 8\end{vmatrix}=0.$$

Thus the vectors $\boldsymbol{a},\boldsymbol{b}$ and $\boldsymbol{c}$ are coplanar and therefore points A,B,C and D lie in the same plane. ■

Example 4.5.6
Let

$$A=\frac{\boldsymbol{b}\times\boldsymbol{c}}{\boldsymbol{a}\cdot\boldsymbol{b}\times\boldsymbol{c}},\ B=\frac{\boldsymbol{c}\times\boldsymbol{a}}{\boldsymbol{a}\cdot\boldsymbol{b}\times\boldsymbol{c}},\ \text{and } C=\frac{\boldsymbol{a}\times\boldsymbol{b}}{\boldsymbol{a}\cdot\boldsymbol{b}\times\boldsymbol{c}}$$

be three vectors in $\mathbf{R}^3$. Show that

(i) $\boldsymbol{A}\cdot\boldsymbol{a}=\boldsymbol{B}\cdot\boldsymbol{b}=\boldsymbol{C}\cdot\boldsymbol{c}=1$;
(ii) $\boldsymbol{A}\cdot\boldsymbol{b}=\boldsymbol{A}\cdot\boldsymbol{c}=\boldsymbol{B}\cdot\boldsymbol{a}=\boldsymbol{B}\cdot\boldsymbol{c}=\boldsymbol{C}\cdot\boldsymbol{a}=\boldsymbol{C}\cdot\boldsymbol{b}=0$.

Solution

(i) $$\boldsymbol{A}\cdot\boldsymbol{a}=\boldsymbol{a}\cdot\boldsymbol{A}=\boldsymbol{a}\cdot\frac{\boldsymbol{b}\times\boldsymbol{c}}{\boldsymbol{a}\cdot\boldsymbol{b}\times\boldsymbol{c}}=\frac{\boldsymbol{a}\cdot\boldsymbol{b}\times\boldsymbol{c}}{\boldsymbol{a}\cdot\boldsymbol{b}\times\boldsymbol{c}}=1;$$

$$\boldsymbol{B}\cdot\boldsymbol{b}=\boldsymbol{b}\cdot\boldsymbol{B}=\boldsymbol{b}\cdot\frac{\boldsymbol{c}\times\boldsymbol{a}}{\boldsymbol{a}\cdot\boldsymbol{b}\times\boldsymbol{c}}=\frac{\boldsymbol{b}\cdot\boldsymbol{c}\times\boldsymbol{a}}{\boldsymbol{a}\cdot\boldsymbol{b}\times\boldsymbol{c}}=\frac{\boldsymbol{a}\cdot\boldsymbol{b}\times\boldsymbol{c}}{\boldsymbol{a}\cdot\boldsymbol{b}\times\boldsymbol{c}}=1;$$

$$\boldsymbol{C}\cdot\boldsymbol{c}=\boldsymbol{c}\cdot\boldsymbol{C}=\boldsymbol{c}\cdot\frac{\boldsymbol{a}\times\boldsymbol{b}}{\boldsymbol{a}\cdot\boldsymbol{b}\times\boldsymbol{c}}=\frac{\boldsymbol{c}\cdot\boldsymbol{a}\times\boldsymbol{b}}{\boldsymbol{a}\cdot\boldsymbol{b}\times\boldsymbol{c}}=\frac{\boldsymbol{a}\cdot\boldsymbol{b}\times\boldsymbol{c}}{\boldsymbol{a}\cdot\boldsymbol{b}\times\boldsymbol{c}}=1;$$

$$\boldsymbol{B}\cdot\boldsymbol{b}=\boldsymbol{b}\cdot\boldsymbol{B}=\boldsymbol{b}\cdot\frac{\boldsymbol{c}\times\boldsymbol{a}}{\boldsymbol{a}\cdot\boldsymbol{b}\times\boldsymbol{c}}=\frac{\boldsymbol{b}\cdot\boldsymbol{c}\times\boldsymbol{a}}{\boldsymbol{a}\cdot\boldsymbol{b}\times\boldsymbol{c}}=\frac{\boldsymbol{a}\cdot\boldsymbol{b}\times\boldsymbol{c}}{\boldsymbol{a}\cdot\boldsymbol{b}\times\boldsymbol{c}}=1;$$

$$\boldsymbol{C}\cdot\boldsymbol{c}=\boldsymbol{c}\cdot\boldsymbol{C}=\boldsymbol{c}\cdot\frac{\boldsymbol{a}\times\boldsymbol{b}}{\boldsymbol{a}\cdot\boldsymbol{b}\times\boldsymbol{c}}=\frac{\boldsymbol{c}\cdot\boldsymbol{a}\times\boldsymbol{b}}{\boldsymbol{a}\cdot\boldsymbol{b}\times\boldsymbol{c}}=\frac{\boldsymbol{a}\cdot\boldsymbol{b}\times\boldsymbol{c}}{\boldsymbol{a}\cdot\boldsymbol{b}\times\boldsymbol{c}}=1;$$

(ii) $$\boldsymbol{A}\cdot\boldsymbol{b}=\boldsymbol{b}\cdot\boldsymbol{A}=\boldsymbol{b}\cdot\frac{\boldsymbol{b}\times\boldsymbol{c}}{\boldsymbol{a}\cdot\boldsymbol{b}\times\boldsymbol{c}}=\frac{\boldsymbol{b}\cdot\boldsymbol{b}\times\boldsymbol{c}}{\boldsymbol{a}\cdot\boldsymbol{b}\times\boldsymbol{c}}=\frac{\boldsymbol{b}\times\boldsymbol{b}\cdot\boldsymbol{c}}{\boldsymbol{a}\cdot\boldsymbol{b}\times\boldsymbol{c}}=0.$$

Similarly, one can prove the rest of (ii). ■

4.6 Triple Cross Product of Vectors

Definition 4.6.1 (Triple cross product)
Let $a, b, c \in \mathbf{R}^3$ be any three vectors. Then the function

$$v : \mathbf{R}^3 \times \mathbf{R}^3 \times \mathbf{R}^3 \rightarrow \mathbf{R}^3,$$

defined by

$$v(a, b, c) = a \times (b \times c) = d$$

is said to be the ***triple cross product.***

Proposition 4.6.1
Let $a, b, c \in \mathbf{R}^3$ be any three vectors. Then,

(i) $(a \times b) \times c = b(a \cdot c) - a(b \cdot c);$
(ii) $a \times (b \times c) = b(a \cdot c) - c(a \cdot b).$

Proof
Before we do an analytic proof, let's try to visualize one of the identities, say (i) (Figure 4.33).

Let

$$d = (a \times b) \times c.$$

Then by definition of the vector product

$$d \perp a \times b \quad \text{and} \quad d \perp c.$$

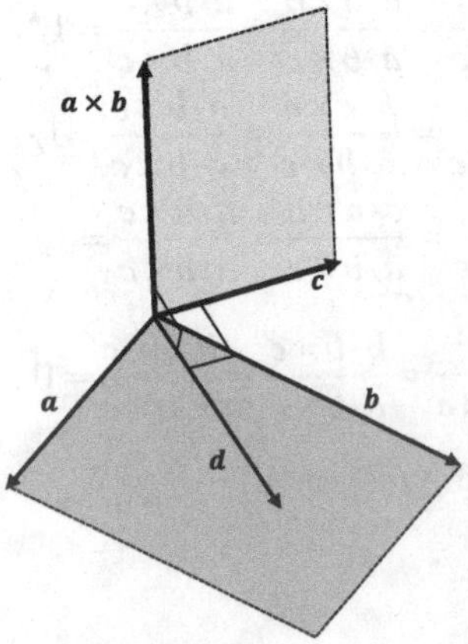

FIGURE 4.33

Now, $\boldsymbol{d}$ being perpendicular to $\boldsymbol{a}\times\boldsymbol{b}$ implies that $\boldsymbol{d}$ lies in the same plane as $\boldsymbol{a}$ and $\boldsymbol{b}$. Therefore, there exists $\alpha,\beta\in\mathbf{R}$ such that

$$\boldsymbol{d}=\alpha\boldsymbol{a}+\beta\boldsymbol{b}.$$

On the other hand, since $\boldsymbol{d}\perp\boldsymbol{c}$,

$$\boldsymbol{d}\perp\boldsymbol{c}=(\alpha\boldsymbol{a}+\beta\boldsymbol{b})\cdot\boldsymbol{c}=\alpha(\boldsymbol{a}\cdot\boldsymbol{c})+\beta(\boldsymbol{b}\cdot\boldsymbol{c})=\mathbf{0}.$$

It follows that

$$\alpha=-\lambda(\boldsymbol{b}\cdot\boldsymbol{c}),$$

$$\beta=\lambda(\boldsymbol{a}\cdot\boldsymbol{c}).$$

The easiest way to prove, say, (ii) analytically is to use the appropriate determinant. So, with

$$\boldsymbol{a}=\alpha_x\boldsymbol{i}+\alpha_y\boldsymbol{j}+\alpha_z\boldsymbol{k},\boldsymbol{b}=\beta_x\boldsymbol{i}+\beta_y\boldsymbol{j}+\beta_z\boldsymbol{k}\,\text{and}\,\boldsymbol{c}=\gamma_x\boldsymbol{i}+\gamma_y\boldsymbol{j}+\gamma_z\boldsymbol{k},$$

we have

$$\begin{aligned}
\boldsymbol{a}\times(\boldsymbol{b}\times\boldsymbol{c})&=(\alpha_x\boldsymbol{i}+\alpha_y\boldsymbol{j}+\alpha_z\boldsymbol{k})\times\begin{vmatrix}\boldsymbol{i}&\boldsymbol{j}&\boldsymbol{k}\\\beta_x&\beta_y&\beta_z\\\gamma_x&\gamma_y&\gamma_z\end{vmatrix}\\
&=(\alpha_x\boldsymbol{i}+\alpha_y\boldsymbol{j}+\alpha_z\boldsymbol{k})\times\left[(\beta_y\gamma_z-\beta_z\gamma_y)\boldsymbol{i}-(\beta_x\gamma_z-\beta_z\gamma_x)\boldsymbol{j}+(\beta_x\gamma_y-\beta_y\gamma_x)\boldsymbol{k}\right]\\
&=\begin{vmatrix}\boldsymbol{i}&\boldsymbol{j}&\boldsymbol{k}\\\alpha_x&\alpha_y&\alpha_z\\\beta_y\gamma_z-\beta_z\gamma_y&\beta_z\gamma_x-\beta_x\gamma_z&\beta_x\gamma_y-\beta_y\gamma_x\end{vmatrix}\\
&=\left[\alpha_y(\beta_x\gamma_y-\beta_y\gamma_x)-\alpha_z(\beta_z\gamma_x-\beta_x\gamma_z)\right]\boldsymbol{i}-\\
&\quad-\left[\alpha_x(\beta_x\gamma_y-\beta_y\gamma_x)-\alpha_z(\beta_y\gamma_z-\beta_z\gamma_y)\right]\boldsymbol{j}+\\
&\quad+\left[\alpha_x(\beta_z\gamma_x-\beta_x\gamma_z)-\alpha_y(\beta_y\gamma_z-\beta_z\gamma_y)\right]\boldsymbol{k}\\
&=(\alpha_y\beta_x\gamma_y-\alpha_y\beta_y\gamma_x-\alpha_z\beta_z\gamma_x+\alpha_z\beta_x\gamma_z)\boldsymbol{i}-\\
&\quad-(\alpha_x\beta_x\gamma_y-\alpha_x\beta_y\gamma_x-\alpha_z\beta_y\gamma_z+\alpha_z\beta_z\gamma_y)\boldsymbol{j}+\\
&\quad+(\alpha_x\beta_z\gamma_x-\alpha_x\beta_x\gamma_z-\alpha_y\beta_y\gamma_z+\alpha_y\beta_z\gamma_y)\boldsymbol{k}.
\end{aligned}\tag{4.9}$$

The right-hand side is

$$
\begin{aligned}
&b(a\cdot c)-c(a\cdot b)=\\
&=(\beta_x i+\beta_y j+\beta_z k)\left[(\alpha_x i+\alpha_y j+\alpha_z k)\cdot(\gamma_x i+\gamma_y j+\gamma_z k)\right]-\\
&\quad-(\gamma_x i+\gamma_y j+\gamma_z k)\left[(\alpha_x i+\alpha_y j+\alpha_z k)\cdot(\beta_x i+\beta_y j+\beta_z k)\right]\\
&=(\beta_x i+\beta_y j+\beta_z k)\cdot(\alpha_x\gamma_x+\alpha_y\gamma_y+\alpha_z\gamma_z)-(\gamma_x i+\gamma_y j+\gamma_z k)\cdot(\alpha_x\beta_x+\alpha_y\beta_y+\alpha_z\beta_z)\\
&=(\alpha_y\beta_x\gamma_y+\alpha_z\beta_x\gamma_z-\alpha_y\beta_y\gamma_x-\alpha_z\beta_z\gamma_x)i+\\
&\quad+(\alpha_x\beta_y\gamma_x+\alpha_z\beta_y\gamma_z-\alpha_x\beta_x\gamma_y-\alpha_z\beta_z\gamma_y)j+\\
&\quad+(\alpha_z\beta_z\gamma_x+\alpha_y\beta_z\gamma_y-\alpha_x\beta_x\gamma_z-\alpha_y\beta_y\gamma_z)k.
\end{aligned}
\tag{4.10}
$$

Comparing (4.9) and (4.10) we conclude that identity (ii) does indeed hold. Similarly, we can prove (i). ■

Corollary 4.6.1 (Jacobi identity)
For any vectors $a,b,c\in\mathbf{R}^3$

$$(a\times b)\times c+(b+c)a+(c\times a)\times b=\mathbf{0}.$$

Proof
From the previous proposition we have

$$(a\times b)\times c=b(a\cdot c)-a(b\cdot c) \tag{4.11}$$

$$(b\times c)\times a=c(b\cdot a)-b(c\cdot a) \tag{4.12}$$

$$(c\times a)\times b=a(c\cdot b)-c(a\cdot b) \tag{4.13}$$

Adding $(1),(2)$ and (3), we get the desired result. ■

Exercise 4.6.1
Show that

(i) $i\times(j\times k)=0;$
(ii) $i\times(j\times i)=-j.$

Exercise 4.6.2
Let $a=(4,-2,1)$, $b=(1,-1,-3)$ and $c=(2,1,2)$. Evaluate

(i) $a\times(b\times c);$
(ii) $(a\times b)\times c.$

Exercise 4.6.3

Let $a = 2i - j - 2k$, $b = 3i + 2k$ and $c = -3i + j + k$.

Show that

(i) $a \times (b \times c) = -21i - 2j - 20k;$

(ii) $(a \times b) \times c = -13i - 7j - 32k.$

Exercise 4.6.4

Show that

(i) there exist $\beta, \gamma \in \mathbf{R}$, such that

$$a \times (b \times c) = \beta b + \gamma c;$$

(ii) there exist $\alpha, \beta \in \mathbf{R}$, such that

$$(a \times b) \times c = \alpha a + \beta b.$$

Example 4.6.1

Let's show that the cross product is not an associative operation, i.e., that

$$(a \times b) \times c \neq a \times (b \times c).$$

Solution

Without loss of generality, we will assume that a, b and c are coplanar and, furthermore, that a and c are not collinear. Then, evidently, $(a \times b) \times c$ and $a \times (b \times c)$ are both coplanar with the plane Π (Figure 4.34(i), (ii)).

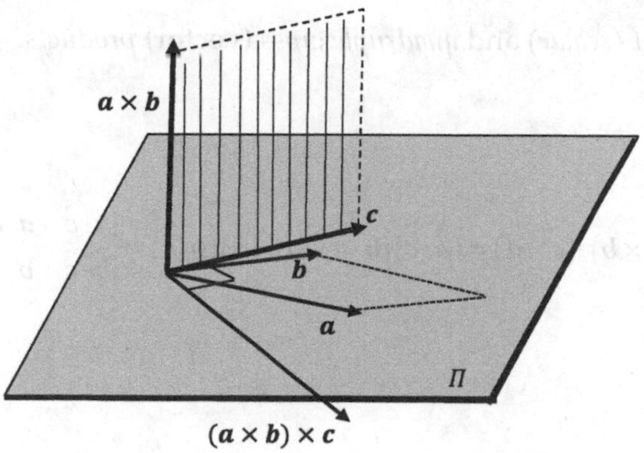

FIGURE 4.34(i)

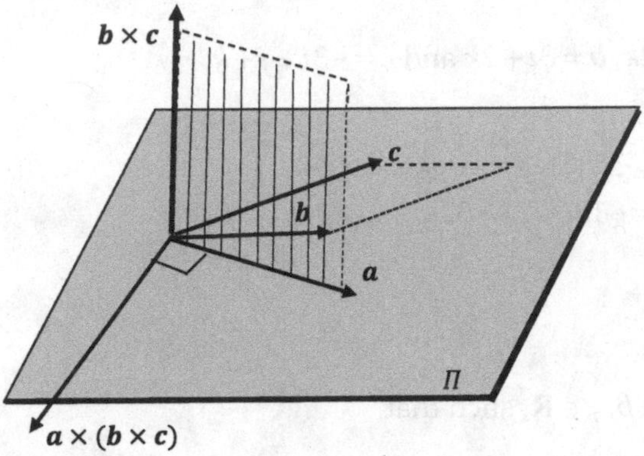

FIGURE 4.34(ii)

Notice, also, that

$$(a \times b) \times c \perp c \quad \text{and} \quad a \times (b \times c) \perp a.$$

Thus, since a and c are not collinear,

$$(a \times b) \times c \neq a \times (b \times c)$$ ■

4.7 The Quadruple Dot and Quadruple Cross Product

Definition 4.7.1 (Quadruple dot and quadruple cross product)
Let $a, b, c, d \in \mathbf{R}^3$ be any four vectors. We call the expressions

$$(a \times b) \cdot (c \times d) \text{and} (a \times b) \times (c \times d).$$

the quadruple dot (*scalar*) and *quadruple cross* **(vector)** *products.*

Example 4.7.1
Show that

$$(a \times b) \cdot (c \times d) = (a \cdot c)(b \cdot d) - (a \cdot d)(b \cdot c) = \begin{vmatrix} a \cdot c & a \cdot d \\ b \cdot c & b \cdot d \end{vmatrix}.$$

Solution
Let $a \times b = v$, then

$$\begin{aligned}(a\times b)\cdot(c\times d) &= v\cdot(c\times d) = (v\times c)\cdot d \\ &= ((a\times b)\times c)\cdot d = (b(a\cdot c) - a(b\cdot c))\cdot d \\ &= (a\cdot c)(b\cdot d) - (a\cdot d)(b\cdot c) \\ &= \begin{vmatrix} a\cdot c & a\cdot d \\ b\cdot c & b\cdot d \end{vmatrix}.\end{aligned}$$

■

Exercise 4.7.1
Show that

$$(a\times b)\cdot(b\times c)\times(c\times a) = (a,b,c)^2.$$

Solution

$$\begin{aligned}(a\times b)\cdot(b\times c)\times(c\times a) &= \begin{vmatrix} a\cdot(b\times c) & a\cdot(c\times a) \\ b\cdot(b\times c) & b\cdot(c\times a) \end{vmatrix} \\ &= \begin{vmatrix} (a,b,c) & 0 \\ 0 & (a,b,c) \end{vmatrix} = (a,b,c)^2\end{aligned}$$

Example 4.7.2
Show that

$$\begin{aligned}(a\times b)\times(c\times d) &= b(a\cdot c\times d) - a(b\cdot c\times d) \\ &= c(a\cdot b\times d) - d(a\cdot b\times c).\end{aligned}$$

Solution
Let $a \times b = v$, then

$$\begin{aligned}(a\times b)\cdot(c\times d) &= v\times(c\times d) = c(v\cdot d) - d(v\cdot c) \\ &= c(a\times b\cdot d) - d(a\times b\cdot c) \\ &= c(a\cdot b\times d) - d(a\cdot b\times c).\end{aligned}$$

On the other hand, with $c \times d = w$, we have

$$\begin{aligned}(a\times b)\times(c\times d) &= (a\times b)\times w = b(a\cdot w) - a(b\cdot w) \\ &= b(a\cdot c\times d) - a(b\cdot c\times d)\end{aligned}$$

Exercise 4.7.2
Prove that

$$\boldsymbol{a}\times\left[\boldsymbol{b}\times(\boldsymbol{c}\times\boldsymbol{d})\right]=\begin{vmatrix}\boldsymbol{a}\times\boldsymbol{c} & \boldsymbol{a}\times\boldsymbol{d}\\ \boldsymbol{b}\cdot\boldsymbol{c} & \boldsymbol{b}\cdot\boldsymbol{d}\end{vmatrix}.$$

Exercise 4.7.3
Simplify

$$\left[(\boldsymbol{a}\times\boldsymbol{b})\times(\boldsymbol{c}\times\boldsymbol{a})\right]\times(\boldsymbol{b}\times\boldsymbol{c}).$$

Notes

1 In advanced linear algebra one distinguishes between vectors written as a row and those written as a column, i.e., between $\boldsymbol{a}=\left(\alpha_x,\alpha_y,\alpha_z\right)$ and $\boldsymbol{a}=\begin{pmatrix}a_x\\ a_y\\ a_z\end{pmatrix}$.

2 Let's stress again the importance of not confusing the coordinates of a point and the components of a vector, i.e. $P(x,y,z)$ is a point whose coordinates are x,y, and z, whereas $\boldsymbol{a}=(x,y,z)=x\boldsymbol{i}+y\boldsymbol{z}+z\boldsymbol{k}$ is a vector. Another useful, and possibly less confusing, notation for a vector via its components is ***a*** = <*x*,*y*,*z*>.

3 A reader not familiar with determinants should consult Appendix C.

5

Elements of Analytic Geometry

5.1 Some Preliminary Remarks

With the tools and techniques developed in the previous sections, we are now in a position to address some topics of analytic geometry. This can also be considered as a simple exercise in the application of a vector algebra. Hence, it might be instructive at this point to remind ourselves of the following: As before, we will be working in the Euclidean space $\mathbf{R}^3$ using as a reference frame the usual Cartesian coordinate system $\{O;(i,j,k)\}$. Recall that we call a Cartesian coordinate system *left-* or *right-handed* depending on the orientation of the basis $\mathcal{B}=\{(i,j,k)\}$. We will continue to use the right-handed system.

Also, we previously used the concept of a *radius vector* without defining it precisely and without emphasizing the equivalence of a space of radius vectors and a space of arbitrary vectors in 3-space. So, here it is more formally.

Consider the set

$$\mathbf{R}^3(O)=\{r_P \mid r_P=\overrightarrow{OP}\}.$$

and a bijection

$$r:\mathbf{R}^3\to\mathbf{R}^3(O)$$

such that for any $a\in\mathbf{R}^3$

$$r(a)=r_P=\overrightarrow{OP}.$$

With the standard operations of the addition and multiplication of vectors, $(\mathbf{R}^3(O);+,\cdot)$ is a vector space.

DOI: 10.1201/9781003343486-5

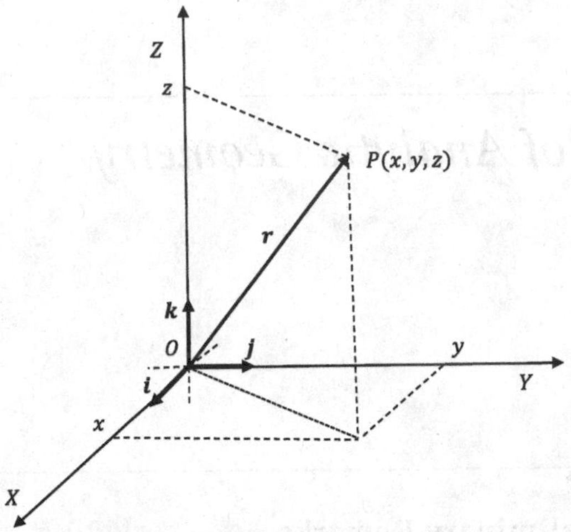

FIGURE 5.1

The simplest orthonormal basis of this space is the familiar set

$$\mathcal{B} = \{\boldsymbol{i} = (1,0,0), \boldsymbol{j} = (0,1,0), \boldsymbol{k} = (0,0,1)\}.$$

Any position vector (Figure 5.1), is therefore uniquely expressed as

$$\boldsymbol{r} = x\boldsymbol{i} + y\boldsymbol{j} + z\boldsymbol{k}, x, y, z \in \mathbf{R}.$$

We say that $x, y, z \in \mathbf{R}$ are the Cartesian coordinates of $\boldsymbol{r}$ in the basis $\mathcal{B}$.

As mentioned before, the advantage of such a definition of a radius vector is that the coordinates of any point P are at the same time the coordinates of the corresponding radius vector. We hope that the reader is already familiar with the statement:

If $A(x_1, y_1, z_1)$ and $B(x_2, y_2, z_2)$ are any two points in $\mathbf{R}^3(O)$, then the vector $\overrightarrow{AB}$ is given by

$$\overrightarrow{AB} = (x_2 - x_1)\boldsymbol{i} + (y_2 - y_1)\boldsymbol{j} + (z_2 - z_1)\boldsymbol{k}.$$

Example 5.1.1

Suppose an object (a particle) is moving from the initial point $A(x_1, y_1, z_1)$ to the final point $B(x_2, y_2, z_2)$, on a trajectory described by a function $f(\boldsymbol{r})$ (Figure 5.2).

The position vectors of the initial point A and the final point B are $\boldsymbol{r}_1 = x_1\boldsymbol{i} + y_1\boldsymbol{j} + z_1\boldsymbol{k}$ and $\boldsymbol{r}_2 = x_2\boldsymbol{i} + y_2\boldsymbol{j} + z_2\boldsymbol{k}$, respectively. The so-called *displacement vector* $\Delta\boldsymbol{r}$ is equal to the difference $\boldsymbol{r}_2 - \boldsymbol{r}_1$, i.e.

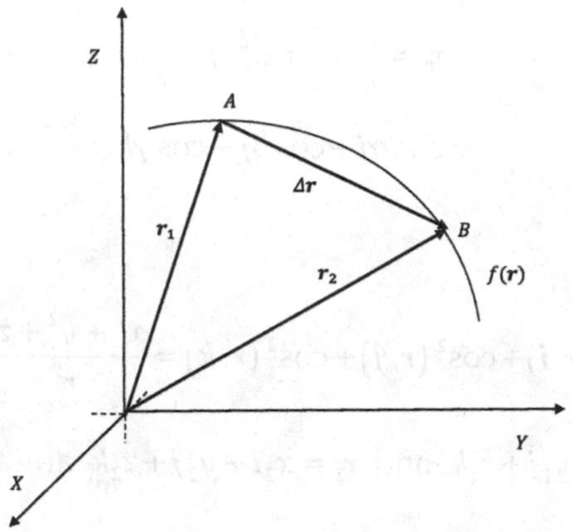

FIGURE 5.2

$$\begin{aligned}\Delta \boldsymbol{r} = \boldsymbol{r}_2 - \boldsymbol{r}_1 &= (x_2\boldsymbol{i} + y_2\boldsymbol{j} + z_2\boldsymbol{k}) - (x_1\boldsymbol{i} + y_1\boldsymbol{j} + z_1\boldsymbol{k}) \\ &= (x_2 - x_1)\boldsymbol{i} + (y_2 - y_1)\boldsymbol{j} + (z_2 - z_1)\boldsymbol{k}.\end{aligned}$$

Then,

$$|\Delta \boldsymbol{r}| = \sqrt{(x_2 - x_1)^2 + (y_2 - y_1)^2 + (z_2 - z_1)^2}$$

represents the distance between points A and B. ■

Let's recall one more thing.

Given a radius vector $\boldsymbol{r} = x\boldsymbol{i} + y\boldsymbol{j} + z\boldsymbol{k}$,

$$\begin{aligned}\boldsymbol{r}\cdot\boldsymbol{i} &= (x\boldsymbol{i} + y\boldsymbol{j} + z\boldsymbol{k})\cdot\boldsymbol{i} \\ &= r \cdot 1 \cdot \cos(\boldsymbol{r}, \boldsymbol{i}) = x.\end{aligned}$$

The *direction cosines* are therefore

$$\cos(\boldsymbol{r}, \boldsymbol{i}) = \cos\alpha = \frac{x}{r},$$

$$\cos(\boldsymbol{r}, \boldsymbol{j}) = \cos\beta = \frac{y}{r},$$

$$\cos(\boldsymbol{r}, \boldsymbol{k}) = \cos\gamma = \frac{z}{r},$$

and the corresponding unit vector is

$$r_0 = \frac{r}{r} = \frac{x}{r}i + \frac{y}{r}j + \frac{z}{r}k$$

$$= \cos\alpha i + \cos\beta j + \cos\gamma k.$$

It follows that

$$\cos^2(r,i) + \cos^2(r,j) + \cos^2(r,k) = \frac{x^2+y^2+z^2}{r^2} = 1.$$

Finally, if $r_1 = x_1 i + y_1 j + z_1 k$ and $r_2 = x_2 i + y_2 j + z_2 k$ are any two (non-zero) radii vectors, then

$$\cos(r_1, r_2) = \frac{r_1 \cdot r_2}{r_1 \cdot r_2} = \frac{x_1 x_2 + y_1 y_2 + z_1 z_2}{r_1 \cdot r_2}$$

$$= \frac{x_1 x_2 + y_1 y_2 + z_1 z_2}{\sqrt{x_1^2 + y_1^2 + z_1^2}\sqrt{x_2^2 + y_2^2 + z_2^2}}.$$

We see immediately that r_1 and r_2 are perpendicular to each other, ($r_1 \perp r_2$), iff $\cos(r_1, r_2) = 0$, or, in other words, iff $x_1 x_2 + y_1 y_2 + z_1 z_2 = 0$ (cf. Corollary 3.2.2).

Example 5.1.2
Suppose the coordinates of the points A and B from the previous example are: $(3,-2,4)$ and $(5,-3,2)$. Therefore, the corresponding radii vectors are $r_1 = 3i - 2j + 4k$ and $r_2 = 5i - 3j + 2k$. Find the direction cosines for the displacement vector Δr.

Solution

$$\Delta r = r_2 - r_1 = (5-3)i + (-3+2)j + (2-4)k$$

$$= 2i - j - 2k.$$

So

$$|\Delta r| = \sqrt{2^2 + (-1)^2 + (-2)^2} = 3.$$

Hence

$$\cos(\Delta r, i) = \frac{x_2 - x_1}{|\Delta r|} = \frac{2}{3},$$

$$\cos(\Delta \boldsymbol{r}, \boldsymbol{j}) = \frac{y_2 - y_1}{|\Delta \boldsymbol{r}|} = -\frac{1}{3},$$

$$\cos(\Delta \boldsymbol{r}, \boldsymbol{k}) = \frac{z_2 - z_1}{|\Delta \boldsymbol{r}|} = -\frac{2}{3}. \quad \blacksquare$$

5.2 Equations of a Line

We start with the familiar general equation of a line in the XY-plane:

$$Ax + By + C = 0. \tag{5.1}$$

If our line passes through the point $P_0(x_0, y_0)$, then the following has to hold

$$Ax_0 + By_0 + C = 0. \tag{5.2}$$

Subtracting (5.2) from (5.1), we obtain the equation of a line l passing through the point P_0:

$$A(x - x_0) + B(y - y_0) + C = 0. \tag{5.3}$$

On a closer inspection of Equation (5.3), we realize that it can be obtained as a dot product of two vectors

$$\boldsymbol{n} = A\boldsymbol{i} + B\boldsymbol{j} \tag{5.4}$$

and

$$\boldsymbol{r} - \boldsymbol{r}_0 = (x - x_0)\boldsymbol{i} + (y - y_0)\boldsymbol{j}.$$

Indeed,

$$\begin{aligned} \boldsymbol{n} \cdot (\boldsymbol{r} - \boldsymbol{r}_0) &= (A\boldsymbol{i} + B\boldsymbol{j}) \cdot ((x - x_0)\boldsymbol{i} + (y - y_0)\boldsymbol{j}) \\ &= A(x - x_0) + B(y - y_0) + C = 0. \end{aligned}$$

So the coefficients A and B are the coordinates of the vector $\boldsymbol{n}$ which is perpendicular to the line l (Figure 5.3(i)).

Any line in either 2- or 3-dimensional (Euclidean) space, $\mathbf{R}^2$ or $\mathbf{R}^3$, is uniquely determined by two points, i.e., there is one and only one line passing through the given points P_1 and P_2.

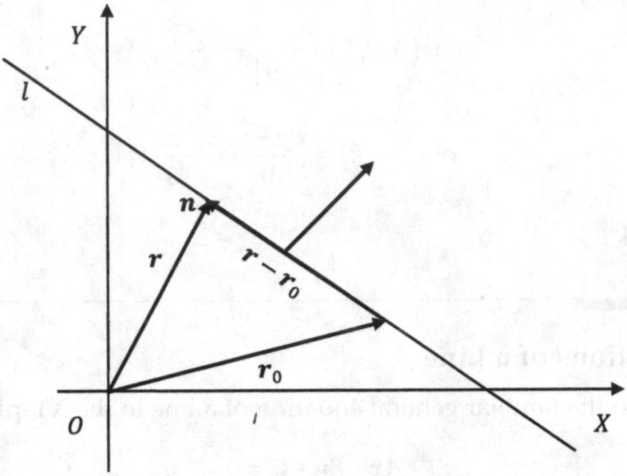

FIGURE 5.3 (i)

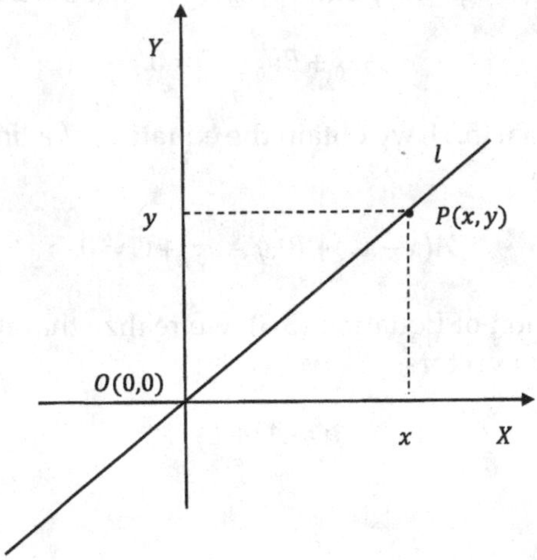

FIGURE 5.3 (ii)

As an example, a line l that passes through the origin of a coordinate system and another point P other than $O(0,0)$ or $O(0,0,0)$ looks something like the line in Figure 5.3(ii) or Figure 5.3(iii).

Let's now consider a line l in a 3-dimensional space $\mathbf{R}^3$, parallel to a given *direction vector* $\boldsymbol{a} = \alpha_x \boldsymbol{i} + \alpha_y \boldsymbol{j} + \alpha_z \boldsymbol{k}$, and passing through a point $P_0(x_0, y_0, z_0)$. Notice that $\boldsymbol{a}$ and $\boldsymbol{a}'$ are isomorphic (Figure 5.4).

Set

$$\overrightarrow{OP_0} = \boldsymbol{r}_0 = x_0 \boldsymbol{i} + y_0 \boldsymbol{j} + z_0 \boldsymbol{k},$$

and let's pick another point, P, on l such that

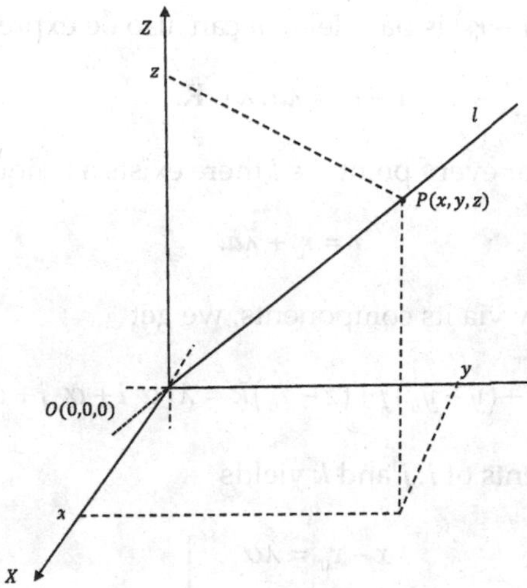

FIGURE 5.3 (iii)

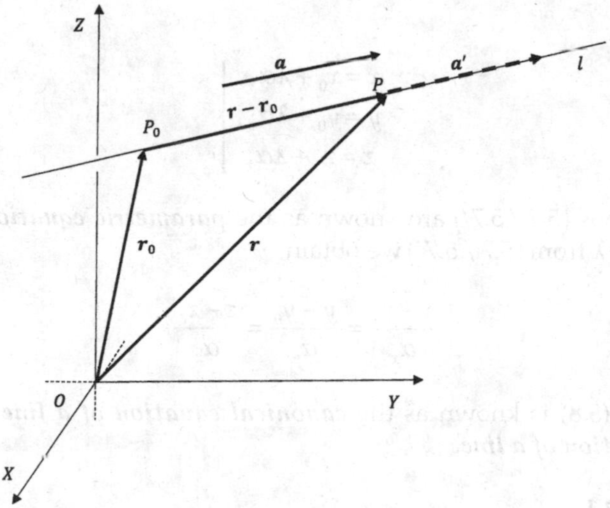

FIGURE 5.4

$$\overrightarrow{OP} = \boldsymbol{r} = x\boldsymbol{i} + y\boldsymbol{j} + z\boldsymbol{k}.$$

Then, by our construction, $\boldsymbol{r} - \boldsymbol{r}_0$ is parallel to $\boldsymbol{a}$, thus

$$(\boldsymbol{r} - \boldsymbol{r}_0) \times \boldsymbol{a} = 0. \tag{5.5}$$

We have obtained the ***vector equation of a line.***

Now, the fact that $(\boldsymbol{r}-\boldsymbol{r}_0)$ is parallel to $\boldsymbol{a}$ can also be expressed by

$$\boldsymbol{r}-\boldsymbol{r}_0=\lambda\boldsymbol{a}, \lambda\in\mathbf{R}. \tag{5.6}$$

To put it differently, for every point $P\in l$ there exists a unique $\lambda\in\mathbf{R}$, such that

$$\boldsymbol{r}=\boldsymbol{r}_0+\lambda\boldsymbol{a}. \tag{5.6'}$$

Writing (5.6) explicitly via its components, we get

$$(x-x_0)\boldsymbol{i}+(y-y_0)\boldsymbol{j}+(z-z_0)\boldsymbol{k}=\lambda(\alpha_x\boldsymbol{i}+\alpha_y\boldsymbol{j}+\alpha_z\boldsymbol{k}).$$

Equating the coefficients of $\boldsymbol{i},\boldsymbol{j}$ and $\boldsymbol{k}$ yields

$$\left.\begin{aligned} x-x_0&=\lambda\alpha_x,\\ y-y_0&=\lambda\alpha_y,\\ z-z_0&=\lambda\alpha_z \end{aligned}\right\} \tag{5.7}$$

or

$$\left.\begin{aligned} x&=x_0+\lambda\alpha_x,\\ y&=y_0+\lambda\alpha_y,\\ z&=z_0+\lambda\alpha_z \end{aligned}\right\} \tag{5.7'}$$

The Equations $(5.7/5.7')$ are known as the ***parametric equations of a line.*** Eliminating λ from $(5.7/5.7')$ we obtain

$$\frac{x-x_0}{\alpha_x}=\frac{y-y_0}{\alpha_y}=\frac{z-z_0}{\alpha_z}. \tag{5.8}$$

Expression (5.8) is known as the ***canonical equation of a line*** or the ***symmetric equation of a line.***

Example 5.2.1
Find the equation of the line l passing through $P_0(4,1,5)$ and parallel to the vector $\boldsymbol{a}=2\boldsymbol{i}-2\boldsymbol{j}+3\boldsymbol{k}$.

Solution
Directly from Equation (5.7) we obtain the parametric equations

$$x=4+2\lambda$$

$$y=1-2\lambda,$$

$$z=5+3\lambda$$

or equivalently, from Equation (5.8) we get the canonical equations for l

$$\frac{x-4}{2}=\frac{y-1}{-2}=\frac{z-5}{3}.$$

■

Example 5.2.2
Find the (canonical) equation of the line passing through the point $P(2,3,4)$ and parallel to the vector $\boldsymbol{a}=5\boldsymbol{i}-\boldsymbol{k}$.

Solution
From Equation (5.8) above, we have

$$\frac{x-2}{5}=\frac{y-3}{0}=\frac{z-4}{-1}. \tag{5.9}$$

Hence, one way to determine the equation of a line is from a point through which it passes and a vector to which it is parallel.

Remark 5.2.1: It is important to note that, in the Equation (5.9) above, the expression $\frac{y-3}{0}$ simply means that, since $\alpha_y=0$, Equation (3') yields $y=3$, and **not** that we are dividing $(y-3)$ by zero.

Example 5.2.3
Find the parametric equations of a line passing through $P_0(1,-3,1)$ and $P(-2,4,5)$.

Solution
The position vectors corresponding to points P_0 and P are:

$$\boldsymbol{r}_0=\overrightarrow{OP_0}=\boldsymbol{i}-3\boldsymbol{j}+\boldsymbol{k} \text{ and } \boldsymbol{r}=\overrightarrow{OP}=-2\boldsymbol{i}+4\boldsymbol{j}+5\boldsymbol{k}.$$

Thus, any vector $\boldsymbol{a}$ in the direction of the line can be expressed as

$$\begin{aligned}\boldsymbol{a}&=\alpha_x\boldsymbol{i}+\alpha_y\boldsymbol{j}+\alpha_z\boldsymbol{k}=\lambda(\boldsymbol{r}-\boldsymbol{r}_0)\\&=\lambda(-3\boldsymbol{i}+7\boldsymbol{j}+4\boldsymbol{k})\end{aligned}$$

Hence

$$x=1-3\lambda,$$

$$y=-3+7\lambda,$$

$$z=1+4\lambda.$$

■

Example 5.2.4
Show that a line l, given by

$$r = r_0 + \lambda a, \tag{5.9}$$

where $r_0 = (12, 28, 0)$ and $a = (9, 21, 0)$, passes through the origin.

Solution
Writing the equation of the line l explicitly as follows

$$(x, y, z) = (12, 28, 0) + \lambda(9, 21, 0),$$

we immediately notice that $l \in \mathbf{R}^2$, i.e. l is in the XY-plane. Therefore, if the line passed through the origin, we would have

$$(0, 0) = (12, 28) + \lambda(9, 21).$$

In other words,

$$0 = 12 + 9\lambda,$$

$$0 = 28 + 21\lambda.$$

Solving both equations for λ we get $\lambda = -\frac{4}{3}$. Thus, from (5.9) it immediately follows that the line l indeed passes through the origin. ■

Exercise 5.2.1
Show that the vector equation of the line passing through the points $A(2, 3, -1)$ and $B(3, 1, 3)$ is

$$r = (2, 3, -1) + \lambda(1, -2, 4).$$

Exercise 5.2.2
Show that the equation of the line passing through the point $A(-2, 5, 1)$ in the direction of the vector $a = (1, -1, 2)$ is

$$r = (-2, 5, 1) + \lambda(1, -1, 2).$$

Exercise 5.2.3
Find the direction cosines for a line passing through the points $A(3, -2, 4)$ and $B(5, -3, 2)$.

Exercise 5.2.4
Find parametric equations of the line l:

(i) passing through the point $P(4,5,2)$ and parallel to $\boldsymbol{a} = 2\boldsymbol{i} - 3\boldsymbol{j} + 3\boldsymbol{k}$;
(ii) passing through the point $P(4,5,2)$ and parallel to $\boldsymbol{a} = 5\boldsymbol{i} - \boldsymbol{k}$.

Exercise 5.2.5
Show that the canonical equation of a line passing through the origin and the point $P(x_1, y_1, z_1)$ is

$$\frac{x}{x_1} = \frac{y}{y_1} = \frac{z}{z_1}.$$

Exercise 5.2.6
Let l be a line passing through the points $P_1(1,0,2)$ and $P_2(3,1,-2)$. Determine whether the point $P(7,3,10)$ lies on the line l.

Exercise 5.2.7
Convince yourself that the point $(3,7)$ does not lie on the line described by

$$(x,y) = (4,-3) + \lambda(11,2).$$

Now let's look for yet another (analogous) way of constructing the equation of a line. Let $A(x_1, y_1, z_1)$ and $B(x_2, y_2, z_2)$ be two distinct points determining the line l, and let $P(x,y,z)$ be another point lying on the line (Figure 5.5), so that

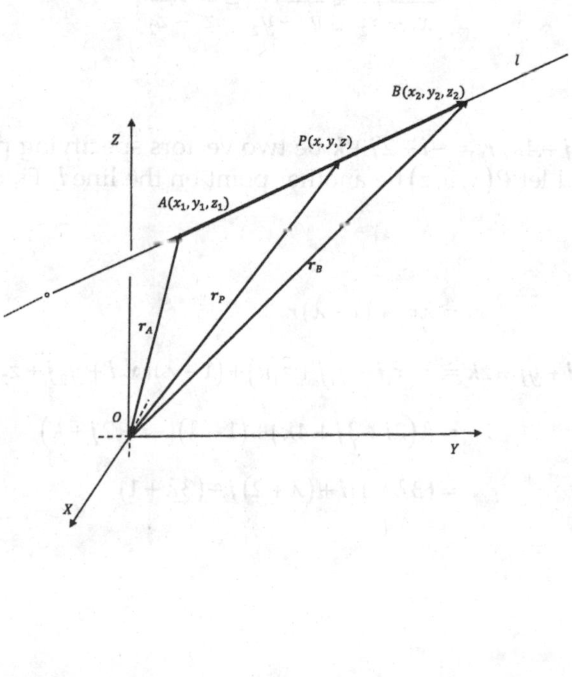

FIGURE 5.5

$$r_A = \overrightarrow{OA},\ r_B = \overrightarrow{OB} \text{ and } r_P = \overrightarrow{OP}.$$

Since $\overrightarrow{AB} = r_B - r_A$,

$$r_P - r_A = \lambda(r_B - r_A),$$

and

$$r_P = (1-\lambda)r_A + \lambda r_B, \quad (5.10)$$

(cf. Corollary 3.4.1).

With $r_P = xi + yj + zk$, $r_A = x_1i + y_1j + z_1k$, and $r_B = x_2i + y_2j + z_2k$ we have

$$xi + yj + zk = (1-\lambda)(x_1i + y_1j + z_1k) + \lambda(x_2i + y_2j + z_2k).$$

Therefore,

$$\left.\begin{aligned} x &= (1-\lambda)x_1 + \lambda x_2 \\ y &= (1-\lambda)y_1 + \lambda y_2 \\ z &= (1-\lambda)z_1 + \lambda z_2 \end{aligned}\right\} \quad (5.11)$$

which is again a set of parametric equations of a line.

Eliminating λ from (5.11) yields the ***two-point form of the equation of a line:***

$$\frac{x-x_2}{x_1-x_2} = \frac{y-y_2}{y_1-y_2} = \frac{z-z_2}{z_1-z_2}. \quad (5.12)$$

Example 5.2.5

Let $r_A = 2i + 3j + 4k$, $r_B = -i + 2j + k$ be two vectors specifying points A and B on a line l, and let $P(x, y, z)$ be another point on the line l. Find the equation of the line l.

Solution

$$r_P = \lambda r_A + (1-\lambda)r_B$$

$$\begin{aligned} xi + yj + zk &= \lambda(x_1i + y_1j + z_1k) + (1-\lambda)(x_2i + y_2j + z_2k) \\ &= \lambda(2i + 3j + 4k) + (1-\lambda)(-i + 2j + k) \\ &= (3\lambda - 1)i + (\lambda + 2)j + (3\lambda + 1) \end{aligned}$$

Thus,

$$x = 3\lambda - 1,$$
$$y = \lambda + 2,$$
$$z = 3\lambda + 1.$$

Solving for λ we get

$$\lambda = \frac{x+1}{3},\ \lambda = \frac{y-2}{1},\ \lambda = \frac{z-1}{3}.$$

Hence

$$\frac{x+1}{3} = \frac{y-2}{1} = \frac{z-1}{3}. \quad \blacksquare$$

Proposition 5.2.1
Let $P_1, P_2, P_3 \in \mathbf{R}^3$ be three distinct points determined by the position vectors $\boldsymbol{r}_1, \boldsymbol{r}_2$ and $\boldsymbol{r}_3$, respectively. Then P_1, P_2, P_3 are collinear iff $\boldsymbol{r}_2 - \boldsymbol{r}_1$ and $\boldsymbol{r}_3 - \boldsymbol{r}_1$ are linearly dependent.

Proof
Suppose that the points are collinear. Then P_3 must lie on the line determined by $\boldsymbol{r}_1$ and $\boldsymbol{r}_2$. Hence there is a scalar λ such that

$$\boldsymbol{r}_3 = \boldsymbol{r}_1 + \lambda(\boldsymbol{r}_2 - \boldsymbol{r}_1)$$

or

$$(\boldsymbol{r}_3 - \boldsymbol{r}_1) - \lambda(\boldsymbol{r}_2 - \boldsymbol{r}_1) = \mathbf{0}.$$

Thus the vectors $\boldsymbol{r}_3 - \boldsymbol{r}_1$ and $\boldsymbol{r}_2 - \boldsymbol{r}_1$ are linearly dependent.

Conversely, suppose there are scalars α_1 and α_2 not both equal to zero, such that

$$\alpha_1(\boldsymbol{r}_2 - \boldsymbol{r}_1) + \alpha_2(\boldsymbol{r}_3 - \boldsymbol{r}_1) = \mathbf{0}.$$

Now, if $\alpha_1 \neq 0$, then

$$(\boldsymbol{r}_2 - \boldsymbol{r}_1) = -\frac{\alpha_2}{\alpha_1}(\boldsymbol{r}_3 - \boldsymbol{r}_1),$$

i.e.

$$\boldsymbol{r}_2 = \boldsymbol{r}_1 + \lambda(\boldsymbol{r}_3 - \boldsymbol{r}_1),$$

where, of course, $\lambda = -\frac{\alpha_2}{\alpha_1}$. Therefore, P_2 lies on the line determined by $r_1$and r_3. In other words, P_1, P_2, P_3 are collinear. The case when $\alpha_2 \neq 0$ can be treated analogously. ■

Corollary 5.2.1
Three distinct points $P_1, P_2, P_3 \in \mathbf{R}^3$, determined by the position vectors r_1, r_2 and r_3 respectively, are collinear iff there exist scalars $\alpha_1, \alpha_2, \alpha_3 \in \mathbf{R}$, not all equal to zero, such that

$$\alpha_1 r_1 + \alpha_2 r_2 + \alpha_3 r_3 = 0$$

with $\alpha_1 + \alpha_2 + \alpha_3 = 0$.

Proposition 3.4.8 can now be restated as

Proposition 5.2.2
If r_1, r_2, r_3 are three non-zero, non-coplanar radius vectors, then any vector in $\mathbf{R}^3$ space can be expressed as a linear combination of r_1, r_2, r_3.

5.3 The Angle Between Two Lines

Definition 5.3.1
Let l_1 and l_2 be two lines in $\mathbf{R}^3$ (Figure 5.6(i)). Then, by the angle

$$\theta = \sphericalangle(l_1, l_2), \quad \theta \leq \frac{\pi}{2},$$

between them, we mean the smaller of the two supplementary angles formed by the two lines l_1', l_2' (Figure 5.6(ii)) parallel to l_1 and l_2 respectively, and intersecting at the point P.

Observe that l_1 and l_2 are two lines given by

$$r = r_1 + \lambda a_1$$

and

$$r = r_2 + \lambda a_2,$$

where

$$r_i = (x_i, y_i, z_i), i = 1, 2.$$

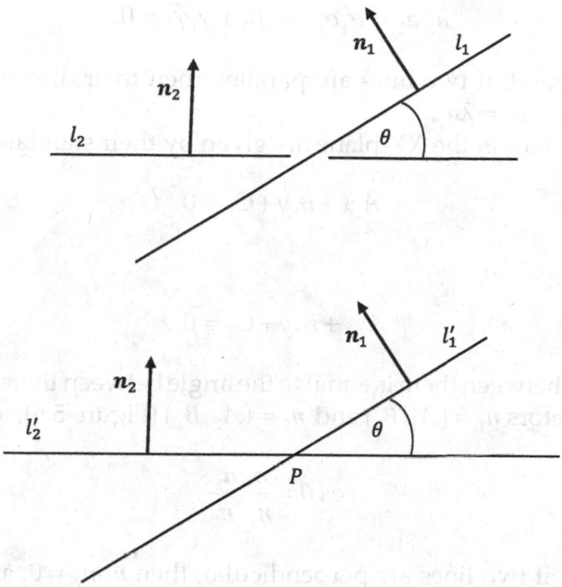

FIGURE 5.6 (i), (ii)

and the corresponding direction vectors are

$$a_i = (\alpha_i, \beta_i, \gamma_i), i = 1, 2.$$

The angle between the direction vectors is

$$\theta = \begin{cases} \measuredangle(a_1, a_2) \text{ if } \measuredangle(a_1, a_2) \le \dfrac{\pi}{2} \\ \pi - \measuredangle(a_1, a_2) \text{ if } \measuredangle(a_1, a_2) > \dfrac{\pi}{2} \end{cases}$$

In either case

$$\cos\theta = |\cos(a_1, a_2)|$$

$$= \frac{|a_1 \cdot a_2|}{a_1 \cdot a_2}$$

$$= \frac{|\alpha_1\alpha_2 + \beta_1\beta_2 + \gamma_1\gamma_2|}{\sqrt{\alpha_1^2 + \beta_1^2 + \gamma_1^2}\sqrt{\alpha_2^2 + \beta_2^2 + \gamma_2^2}}.$$

So, again, we see that two lines are perpendicular if the scalar product of their direction vectors is equal to zero, i.e.

$$\boldsymbol{a}_1 \cdot \boldsymbol{a}_2 = \alpha_1\alpha_2 + \beta_1\beta_2 + \gamma_1\gamma_2 = 0.$$

On the other hand, if two lines are parallel, then their direction vectors are proportional, i.e. $\boldsymbol{a}_1 = \lambda \boldsymbol{a}_2$.

Now, if two lines in the *XY*-plane are given by their standard equations

$$A_1x + B_1y + C_1 = 0$$

and

$$A_2x + B_2y + C_2 = 0,$$

then the angle between them is equal to the angle between their two respective orthogonal vectors $\boldsymbol{n}_1 = (A_1, B_1)$ and $\boldsymbol{n}_2 = (A_2, B_2)$ (Figure 5.6), and

$$\cos\theta = \frac{\boldsymbol{n}_1 \cdot \boldsymbol{n}_2}{n_1 \cdot n_2}.$$

Consequently, if two lines are perpendicular, then $\boldsymbol{n}_1 \cdot \boldsymbol{n}_2 = 0$, and if two lines are parallel, then $\boldsymbol{n}_1 = \lambda \boldsymbol{n}_2$.

Example 5.3.1
Find the cosine of the angle between the two lines l_1 and l_2 given by:

$$l_1 : 3x - 4y + 1 = 0$$

$$l_2 : 2x + y - 5 = 0$$

Solution

$$\cos\theta = \frac{\boldsymbol{n}_1 \cdot \boldsymbol{n}_2}{n_1 \cdot n_2} = \frac{3 \cdot 2 + (-4) \cdot 1}{\sqrt{3^2 + (-4)^2}\sqrt{2^2 + 1^2}} = \frac{2}{5\sqrt{5}}.$$ ■

Exercise 5.3.1
Find the angle between the lines

$$l_1 : \frac{x-1}{1} = \frac{y}{2} = \frac{z-5}{-2} \quad \text{and} \quad l_2 : \frac{x+6}{3} = \frac{y-2}{-4} = \frac{z}{5}.$$

Definition 5.3.2
Let l be a line whose direction vector is $\boldsymbol{a}$, and let Π be a plane in $\mathbf{R}^3$. By the angle $\theta = \sphericalangle(l, \Pi)$ between l and Π we mean the angle between l and its orthogonal projection l' on the plane Π (Figure 5.7).

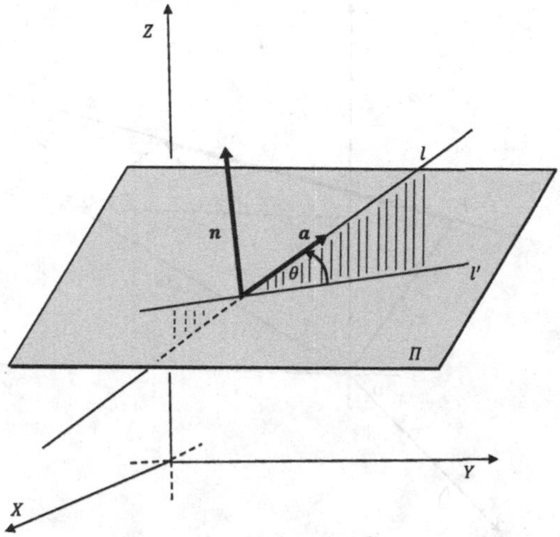

FIGURE 5.7

It follows that for the angle between two lines l and l' we have

$$\cos\left(\frac{\pi}{2}-\theta\right)=|\cos(a,n)|=\frac{|a\cdot n|}{|a|\cdot|n|}.$$

In other words,

$$\sin\theta=\frac{|a\cdot n|}{|a|\cdot|n|}.$$

5.4 The Distance Between a Point and a Line

Definition 5.4.1
Let l be a line in $\mathbf{R}^3$. By the distance $d=d(P_0,l)$ between a point P_0 and a line l we mean the length of the normal drawn from the point P_0 to the line l (Figure 5.8).

So, to find the distance $d(P_0,l)$ between a line l given by

$$r=r_1+\lambda \boldsymbol{l},$$

where $\boldsymbol{l}=\overrightarrow{P_1P_2}$, and a point P_0, we reason as follows:

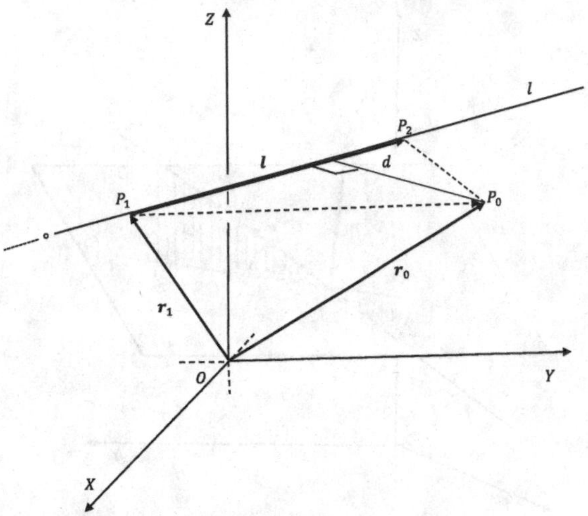

FIGURE 5.8

The area A of the triangle $\Delta P_0P_1P_2$ is

$$A = \frac{1}{2}(l \cdot d).$$

The same area is also given by

$$A = \frac{1}{2}\left|\left(\overrightarrow{P_1P_0} \times \overrightarrow{P_1P_2}\right)\right|$$

$$= \frac{1}{2}\left|(\boldsymbol{r}_0 - \boldsymbol{r}_1) \times \boldsymbol{l}\right|.$$

Thus,

$$\frac{1}{2}(l \cdot d) = \frac{1}{2}\left|(\boldsymbol{r}_0 - \boldsymbol{r}_1) \times \boldsymbol{l}\right|,$$

and therefore

$$d = \frac{\left|(\boldsymbol{r}_0 - \boldsymbol{r}_1) \times \boldsymbol{l}\right|}{l}. \tag{5.13}$$

We can approach this in yet another way:

Let

$$\boldsymbol{r}_0 = x_0\boldsymbol{i} + y_0\boldsymbol{j} + z_0\boldsymbol{k},$$

$$r_1 = x_1 i + y_1 j + z_1 k,$$

and

$$l = \alpha i + \beta j + \gamma k,$$

be as in Figure 5.8.

Then we have

$$r_0 - r_1 = (x_0 - x_1)i + (y_0 - y_1)j + (z_0 - z_1)k.$$

So

$$d = \frac{|[(x_0 - x_1)i + (y_0 - y_1)j + (z_0 - z_1)k] \times (\alpha i + \beta j + \gamma k)|}{l}$$

$$= \frac{1}{l}\left\| \begin{matrix} i & j & k \\ x_0 - x_1 & y_0 - y_1 & z_0 - z_1 \\ \alpha & \beta & \gamma \end{matrix} \right\|. \qquad (5.14)$$

Exercise 5.4.1
Show that (5.13) and (5.14) in Definition 5.4.1 are equivalent.

Example 5.4.1
Find the distance between the point $P_0(2,3,1)$ and the line given by

$$\frac{x-1}{1} = \frac{y+2}{2} = \frac{z-2}{-2}.$$

Solution
First, observe that $r_0 = 2i + 3j + k, r_1 = i - 2j + 2k$, and, since $l = i + 2j - 2k$, $l = 3$.

Now, to simplify our calculation of (5.13) let's introduce the following notation

$$X = \begin{vmatrix} y_0 - y_1 & z_0 - z_1 \\ \beta & \gamma \end{vmatrix}, Y = -\begin{vmatrix} x_0 - x_1 & z_0 - z_1 \\ \alpha & \gamma \end{vmatrix}, Z = \begin{vmatrix} x_0 - x_1 & y_0 - y_1 \\ \alpha & \beta \end{vmatrix}.$$

Then we have

$$d = \frac{1}{l}\sqrt{X^2 + Y^2 + Z^2}.$$

Since

$$X = \begin{vmatrix} 5 & -1 \\ 2 & -2 \end{vmatrix} = -8, \quad Y = -\begin{vmatrix} 1 & -1 \\ 1 & -2 \end{vmatrix} = 1, \quad Z = \begin{vmatrix} 1 & 5 \\ 1 & 2 \end{vmatrix} = -3,$$

the distance we are looking for is

$$d = \frac{1}{3}\sqrt{74}$$

■

One more way to find the distance from a point to a line is as follows.

Consider a line l given by $Ax + By + C = 0$, and a point $P_1(x_1, y_1)$ not on the line (Figure 5.9).

If $P_0(x_0, y_0)$ is any point on the line l, then

$$Ax_0 + By_0 + C = 0.$$

Furthermore, let $\boldsymbol{n} = A\boldsymbol{i} + B\boldsymbol{j}$ be a vector normal to l. Then by the distance d between a point P_1 and a line l we mean the shortest distance between P_1 and the line l, i.e., the length of the segment $\overline{P_1P_1'}$ of the line perpendicular to l. But the distance d is also the magnitude of the projection of $\boldsymbol{r} = \overrightarrow{P_0P_1}$ on $\boldsymbol{n}$, i.e.

$$d = \frac{\left|\overrightarrow{P_0P_1} \cdot \boldsymbol{n}\right|}{|\boldsymbol{n}|} = \frac{|\boldsymbol{r} \cdot \boldsymbol{n}|}{n}.$$

So, with

$$\overrightarrow{OP_0} = \boldsymbol{r}_0 = x_0\boldsymbol{i} + y_0\boldsymbol{j}$$

and

$$\overrightarrow{OP_1} = \boldsymbol{r}_1 = x_1\boldsymbol{i} + y_1\boldsymbol{j},$$

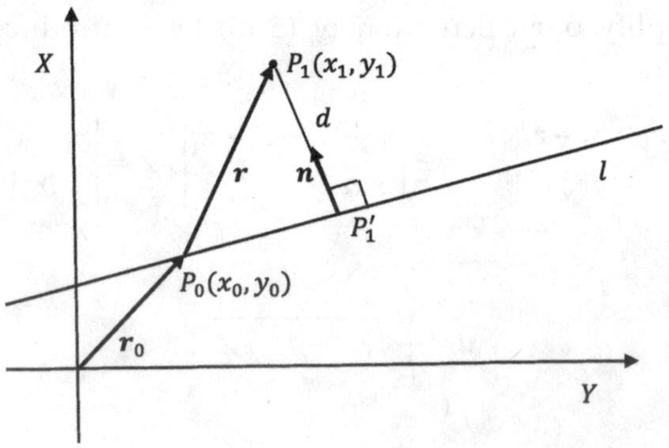

FIGURE 5.9

$$r = (x_1 - x_0)i + (y_1 - y_0)j.$$

Furthermore, since

$$n = Ai + Bj,$$

and

$$C = -Ax_0 - By_0,$$

we finally have

$$d = \frac{|A(x_1 - x_0) + B(y_1 - y_0)|}{\sqrt{A^2 + B^2}}$$

$$= \frac{|Ax_1 + By_1 - Ax_0 - By_0|}{\sqrt{A^2 + B^2}}$$

$$= \frac{|Ax_1 + By_1 + C|}{\sqrt{A^2 + B^2}}.$$

■

Example 5.4.2
Find the distance from $P_1(3,2)$ to the line $l: 3x + 4y - 7 = 0$.

Solution

$$d = \frac{|Ax_1 + By_1 + C|}{\sqrt{A^2 + B^2}} = \frac{|3 \cdot 3 + 4 \cdot 2 - 7|}{\sqrt{3^2 + 4^2}} = 2.$$

■

Example 5.4.3
Let ΔABC be a triangle in the XY-plane with vertices $A(2,-1), B(4,4), C(9,7)$. Find the altitude h_A drawn from vertex A.

Solution
The altitude h_A equals the distance between the point A and the line l_{BC} passing through the points B and C. So we first find the equation of the line using the two-point form of the equation of a line

$$l_{BC}: \quad \frac{x-4}{4-9} = \frac{y-7}{4-7}.$$

Hence $l_{BC}: 3x - 5y + 8 = 0$, and a vector normal to l_{BC} is $n = (3, -5)$. The distance we are looking for is

$$d=\frac{|Ax_1+By_1+C|}{\sqrt{A^2+B^2}}=\frac{3\cdot 2+(-5)\cdot(-1)+8}{\sqrt{3^2+(-5)^2}}=\frac{19}{\sqrt{29}}. \qquad \blacksquare$$

5.5 The Equations of a Plane

Let Π be a plane in 3-dimensional space, determined by three non-collinear points $P_1(x_1,y_1,z_1)$, $P_2(x_2,y_2,z_2)$, and $P_3(x_3,y_3,z_3)$, and let $\{O;(\mathbf{i},\mathbf{j},\mathbf{k})\}$ be our chosen Cartesian coordinate system. Furthermore, let $\boldsymbol{r}_1,\boldsymbol{r}_2$ and $\boldsymbol{r}_3$ be the corresponding position vectors (Figure 5.10).

If P is any other point lying in the plane Π, then we can construct three vectors

$$\boldsymbol{a}=\overrightarrow{P_1P_2}=\boldsymbol{r}_2-\boldsymbol{r}_1,$$

$$\boldsymbol{b}=\overrightarrow{P_1P_3}=\boldsymbol{r}_3-\boldsymbol{r}_1,$$

$$\boldsymbol{c}=\overrightarrow{P_1P}=\boldsymbol{r}-\boldsymbol{r}_1,$$

which all lie in the plane Π. Invoking Proposition 4.5.2 we write the (vector) equation of the plane as

$$\boldsymbol{c}\cdot(\boldsymbol{a}\times\boldsymbol{b})=\begin{vmatrix} x-x_1 & y-y_2 & z-z_3 \\ x_2-x_1 & y_2-y_1 & z_2-z_1 \\ x_3-x_1 & y_3-y_1 & z_3-z_1 \end{vmatrix}=0. \qquad (5.15)$$

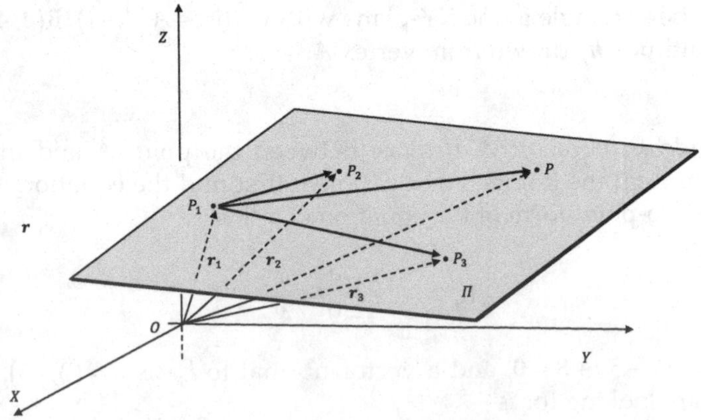

FIGURE 5.10

Example 5.5.1

Find the equation of the plane Π determined by the three points $P_1(1,2,0), P_2(3,-1,2)$ and $P_3(2,4,3)$.

Solution

Consider

$$\overrightarrow{P_1P_2} = r_2 - r_1 = 2i - 3j + 2k,$$

$$\overrightarrow{P_1P_3} = r_3 - r_1 = i + 2j + 3k,$$

and let

$$r = xi + yj + zk$$

be the radius vector of an arbitrary point P also in the plane Π.

Then, the vectors $r, \overrightarrow{P_1P_2}$ and $\overrightarrow{P_1P_3}$ are coplanar and, according to (5.15) above, the equation we are looking for is

$$\begin{vmatrix} x - x_1 & y - y_2 & z - z_3 \\ x_2 - x_1 & y_2 - y_1 & z_2 - z_1 \\ x_3 - x_1 & y_3 - y_1 & z_3 - z_1 \end{vmatrix} = \begin{vmatrix} x-1 & y-2 & z \\ 2 & -3 & 2 \\ 1 & 2 & 3 \end{vmatrix} = 0$$

i.e.

$$13x + 4y - 7z - 21 = 0 \tag{5.16}$$

■

Example 5.5.2

Find the equation of the plane through the points P_1, P_2 and P_3 whose position vectors are $r_1 = 2i - k, r_2 = 3i + 2j + k$ and $r_3 = -i + 4j + 2k$, respectively.

Solution

$$\begin{vmatrix} x - x_1 & y - y_2 & z - z_3 \\ x_2 - x_1 & y_2 - y_1 & z_2 - z_1 \\ x_3 - x_1 & y_3 - y_1 & z_3 - z_1 \end{vmatrix} = \begin{vmatrix} x-2 & y & z+1 \\ 1 & 2 & 2 \\ -3 & 4 & 3 \end{vmatrix} = 0$$

i.e.

$$2x + 9y - 10z - 14 = 0 \tag{5.17}$$

■

Equations (5.16) and (5.17) from the previous two examples are in the form

$$Ax + By + Cz + D = 0,$$

which is known as the ***general equation of a plane.***

Exercise 5.5.1
Show that the equation of a plane passing through $P_1(2,-1,1)$, $P_2(3,2,-1)$ and $P_3(-1,3,2)$ is

$$11x+5y+13z-30=0.$$

Example 5.5.3
Determine whether the points $P_1\left(2,1,\frac{9}{2}\right)$ and $P_2(0,9,-1)$ lie in the plane

$$\Pi: \begin{vmatrix} x-1 & y-2 & z \\ 1 & -1 & 4 \\ 1 & 2 & -3 \end{vmatrix} = 0.$$

Solution
In order for the points $P_1\left(2,1,\frac{9}{2}\right)$ and $P_2(0,9,-1)$ to be coplanar the following has to be satisfied:

$$\begin{vmatrix} x-1 & y-2 & z \\ 1 & -1 & 4 \\ 1 & 2 & -3 \end{vmatrix} = -5x+7y+3z-9=0.$$

But,

$$-5\cdot 2+7\cdot 1+3\cdot\frac{9}{2}-9\neq 0$$

and

$$-5\cdot 0+7\cdot 9+3\cdot(-1)-9\neq 0.$$

Thus, neither P_1 nor P_2 lies in Π. ■

In addition to being uniquely determined by three non-collinear points, a plane in 3-space can be determined by the unit vector $\boldsymbol{n}_0$ perpendicular to the plane Π and the plane's distance from the origin of the coordinate system δ. So, before deriving this equation let's see how we can find the distance $d(P,\Pi)$ between a point P and a plane Π (Figure 5.11).

Suppose that our plane Π is fixed by the unit vector $\boldsymbol{n}_0$ perpendicular to the plane and the plane's distance δ from the origin of the coordinate system $\{O;\boldsymbol{i},\boldsymbol{j},\boldsymbol{k}\}$. Furthermore, let P be some point in space whose position vector is $\boldsymbol{r}$. Suppose we want to find the distance $d(P,\Pi)$ between the point P and the plane Π. Observe that

$$\boldsymbol{r}\cdot\boldsymbol{n}_0 = rn_0\cos(\boldsymbol{r},\boldsymbol{n}_0) = r\cos(\boldsymbol{r},\boldsymbol{n}_0) = \delta+d,$$

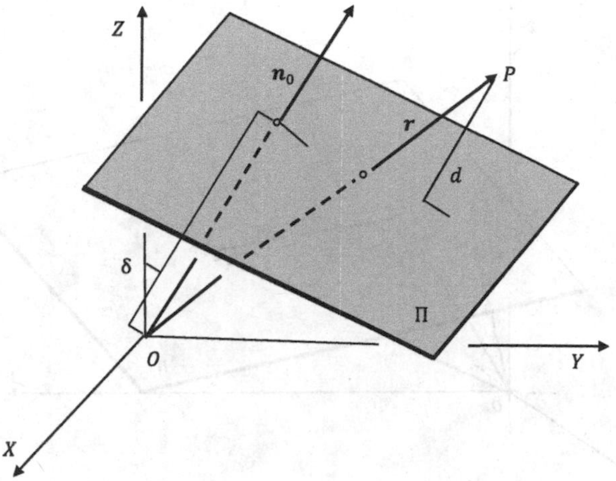

FIGURE 5.11

therefore

$$d = \boldsymbol{r} \cdot \boldsymbol{n}_0 - \delta. \tag{5.18}$$

If

$$\boldsymbol{n}_0 = \cos\vartheta \boldsymbol{i} + \cos\varphi \boldsymbol{j} + \cos\psi \boldsymbol{k}$$

and

$$\boldsymbol{r} = x\boldsymbol{i} + y\boldsymbol{j} + z\boldsymbol{k},$$

then

$$\begin{aligned} d &= (\cos\vartheta \boldsymbol{i} + \cos\varphi \boldsymbol{j} + \cos\psi \boldsymbol{k}) \cdot (x\boldsymbol{i} + y\boldsymbol{j} + z\boldsymbol{k}) - \delta \\ &= x\cos\vartheta + y\cos\varphi + z\cos\psi - \delta. \end{aligned} \tag{5.19}$$

Remark 5.5.1

Ordinarily we would consider "distance" to be a positive quantity, but the distance d could be "negative" in the case when the points P and O are at the opposite sides of the plane Π.

Now we can use this result to construct another equation of a plane. We argue as follows: Let Π be a plane in 3-space determined by the unit vector perpendicular to it and the distance δ from the origin of the coordinate system $\{O; \boldsymbol{i}, \boldsymbol{j}, \boldsymbol{k}\}$. From (1) and (2) it follows that a point P is in the plane iff

$$\boldsymbol{r} \cdot \boldsymbol{n}_0 - \delta = 0 \tag{5.20}$$

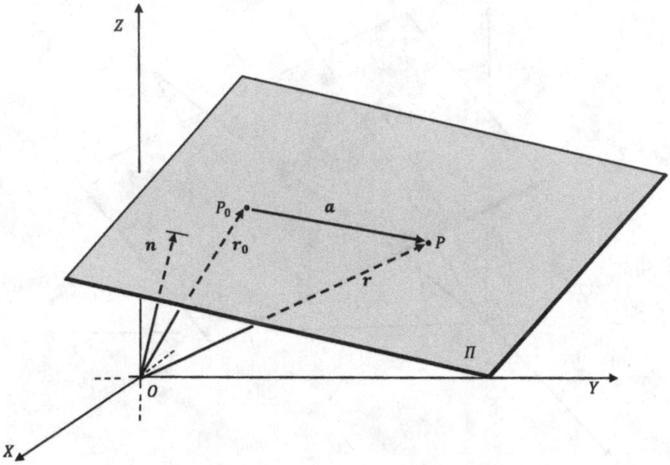

FIGURE 5.12

or

$$x\cos\vartheta + y\cos\varphi + z\cos\psi - \delta = 0 \qquad (5.20')$$

Equations (5.20)/(5.20′) are known as the ***normal*** or the ***Hesse***[1] ***equation of a plane.***

One more way to construct the equation of a plane is possible if one knows the coordinates of two points $P_0(x_0, y_0, z_0)$ and $P(x, y, z)$ lying in the plane Π and the vector $\boldsymbol{n}$ normal to the plane (Figure 5.12).

The coordinates of the points P_0 and P are also the coordinates of the respective position vectors, namely,

$$\overrightarrow{OP_0} = \boldsymbol{r}_0 = x_0\boldsymbol{i} + y_0\boldsymbol{j} + z_0\boldsymbol{k} \text{ and } \overrightarrow{OP} = \boldsymbol{r} = x\boldsymbol{i} + y\boldsymbol{j} + z\boldsymbol{k}.$$

Consequently, the vector

$$\begin{aligned}\boldsymbol{a} &= \overrightarrow{P_0P} = \boldsymbol{r} - \boldsymbol{r}_0 \\ &= (x - x_0)\boldsymbol{i} + (y - y_0)\boldsymbol{j} + (z - z_0)\boldsymbol{k}.\end{aligned}$$

Since $\boldsymbol{n}$ is perpendicular to Π it follows that

$$\boldsymbol{a} \cdot \boldsymbol{n} = 0.$$

Now, if $\boldsymbol{n} = n_1\boldsymbol{i} + n_2\boldsymbol{j} + n_3\boldsymbol{k}$, then

$$\begin{aligned} a\cdot n &= \left[(x-x_0)i+(y-y_0)j+(z-z_0)k\right]\cdot(n_1 i+n_2 j+n_3 k) \\ &= (x-x_0)n_1+(y-y_0)n_2+(z-z_0)n_3 \\ &= (xn_1+yn_2+zn_3)-(x_0n_1+y_0n_2+z_0n_3) \\ &= xn_1+yn_2+zn_3-\delta=0, \end{aligned} \tag{5.21}$$

where $\delta = x_0n_1+y_0n_2+z_0n_3$.

We recognize the analogue of Equation (5.20′). Rewriting Equation (5.21) as

$$xn_1+yn_2+zn_3=\delta, \tag{5.22}$$

and dividing both sides of (5.21) by δ we get

$$\frac{xn_1}{\delta}+\frac{yn_2}{\delta}+\frac{zn_3}{\delta}=1$$

or

$$\frac{x}{\frac{\delta}{n_1}}+\frac{y}{\frac{\delta}{n_2}}+\frac{z}{\frac{\delta}{n_3}}=1.$$

Finally, with

$$\alpha=\frac{\delta}{n_1},\ \beta=\frac{\delta}{n_2},\ \gamma=\frac{\delta}{n_3},$$

we obtain

$$\frac{x}{\alpha}+\frac{y}{\beta}+\frac{z}{\gamma}=1.$$

This is the ***intercept form of the equation of a plane.*** α,β and γ are the x-, y- and z-intercepts of the plane Π (Figure 5.13).

Example 5.5.4
Find the equation of the plane Π through the point $P_0(-2,1,2)$ and normal to the vector $n=-2i+j+2k$.

Solution
Let $r=xi+yj+zk$ be the position vector of any point $P\neq P_0$ in the plane. Consider

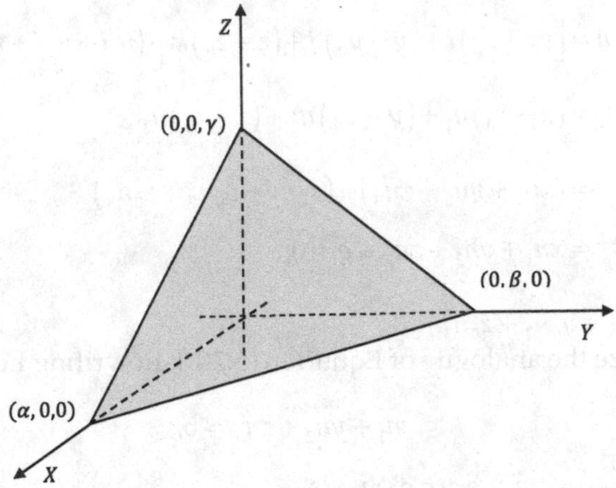

FIGURE 5.13

$$a = \overrightarrow{P_0P} = r - n$$
$$= (xi + yj + zk) - (-2i + j + 2k)$$
$$= (x+2)i + (y-1)j + (z-2)k.$$

Since n and a are perpendicular to each other,

$$n \cdot a = n \cdot (r - n) = 0.$$

So we have

$$n \cdot a = n \cdot (r - n)$$
$$= (-2i + j + 2k) \cdot [(x+2)i + (y-1)j + (z-2)k]$$
$$= -2(x+2) + (y-1) + 2(z-2) = 0$$

Hence the equation of the plane Π is

$$2x - y - 2z + 9 = 0. \qquad \blacksquare$$

Exercise 5.5.2
Show that the equation of the plane Π passing through the point $P_0(1,-2,3)$ and perpendicular to the vector $n = 4i + 4j - 6k$ is $4x + 5y - 6z + 22 = 0$.

Exercise 5.5.3
Find the equation of the plane Π perpendicular to the vector $n = i + j + 3k$ and passing through the point P_0 whose position vector is $a = 2i - j - 2k$.

Example 5.5.5
Find the equation of the plane Π passing through the point $P(4,2,1)$ and parallel to the plane $\Pi_{||}: 2x+3y-z+5=0$.

Solution
Since Π and $\Pi_{||}$ are parallel, the vector $\boldsymbol{n}=(2,3,-1)$ normal to Π is also normal to $\Pi_{||}$. Therefore

$$\delta = 2x+3y-z$$

Since $P(4,2,1)$ lies in the plane,

$$\delta = 2\cdot 4+3\cdot 2-1=13$$

Hence,

$$\Pi: 2x+3y-z-13=0.$$ ■

Next, let's consider a plane in space determined by a point P_0 and two non-collinear vectors $\boldsymbol{a}$ and $\boldsymbol{b}$ (Figure 5.14).

If P is any other point in Π, then the three vectors $\boldsymbol{a},\boldsymbol{b}$ and $\overrightarrow{P_0P}$ are coplanar, and therefore, according to Proposition 3.4.5, there exist two scalars $\lambda,\mu \in \mathbf{R}$, such that

$$\overrightarrow{P_0P} = \lambda \boldsymbol{a} + \mu \boldsymbol{b} \tag{5.23}$$

But, as is evident from the figure,

$$\overrightarrow{P_0P} = \boldsymbol{r} - \boldsymbol{r}_0$$

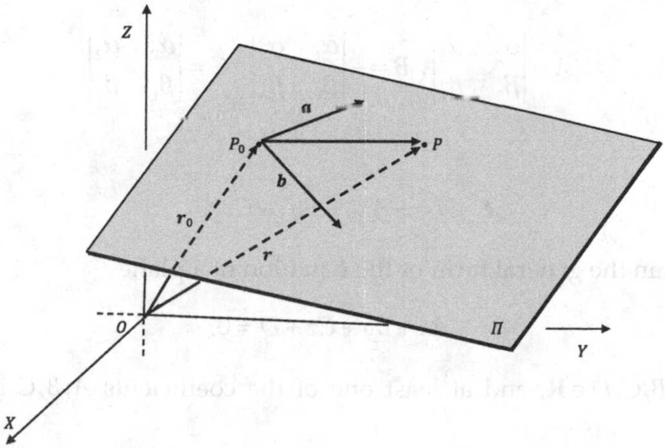

FIGURE 5.14

where, as usual, we take $\boldsymbol{r} = x\boldsymbol{i} + y\boldsymbol{j} + z\boldsymbol{k}$ and $\boldsymbol{r}_0 = x_0\boldsymbol{i} + y_0\boldsymbol{j} + z_0\boldsymbol{k}$.
So,

$$\boldsymbol{r} = \boldsymbol{r}_0 + \lambda \boldsymbol{a} + \mu \boldsymbol{b} \tag{5.24}$$

If $\boldsymbol{a} = \alpha_1\boldsymbol{i} + \alpha_2\boldsymbol{j} + \alpha_3\boldsymbol{k}$ and $\boldsymbol{b} = \beta_1\boldsymbol{i} + \beta_2\boldsymbol{j} + \beta_3\boldsymbol{k}$, then from (5.24) we obtain the ***parametric equations of a plane:***

$$\begin{aligned} x &= x_0 + \lambda\alpha_1 + \mu\beta_1 \\ y &= y_0 + \lambda\alpha_2 + \mu\beta_2 \\ z &= z_0 + \lambda\alpha_3 + \mu\beta_3 \end{aligned} \tag{5.25}$$

In order to get just one equation of a plane, we need to eliminate the parameters λ and μ from the system (5.24). Since, by our construction, vectors $\boldsymbol{a}, \boldsymbol{b}$ and $\boldsymbol{r} - \boldsymbol{r}_0$ are coplanar, their mixed product has to be zero, i.e.

$$\begin{vmatrix} x - x_0 & y - y_0 & z - z_0 \\ \alpha_1 & \alpha_2 & \alpha_3 \\ \beta_1 & \beta_2 & \beta_3 \end{vmatrix} = 0. \tag{5.26}$$

(5.25) is the ***equation of a plane determined by a point and two non-collinear vectors.***

Evaluating the determinant (5.26) we get

$$(x - x_0)\begin{vmatrix} \alpha_2 & \alpha_3 \\ \beta_2 & \beta_3 \end{vmatrix} - (y - y_0)\begin{vmatrix} \alpha_1 & \alpha_3 \\ \beta_1 & \beta_3 \end{vmatrix} + (z - z_0)\begin{vmatrix} \alpha_2 & \alpha_2 \\ \beta_1 & \beta_2 \end{vmatrix} = 0.$$

With

$$A = \begin{vmatrix} \alpha_2 & \alpha_3 \\ \beta_2 & \beta_3 \end{vmatrix}, \; B = -\begin{vmatrix} \alpha_1 & \alpha_3 \\ \beta_1 & \beta_3 \end{vmatrix}, \; C = \begin{vmatrix} \alpha_2 & \alpha_2 \\ \beta_1 & \beta_2 \end{vmatrix}$$

and

$$D = -Ax_0 - By_0 - Cz_0,$$

we get again the general form of the equation of a plane

$$Ax + By + Cz + D = 0, \tag{5.27}$$

where $A, B, C, D \in \mathbf{R}$, and at least one of the coefficients A, B, C is different from zero.

In conclusion, regardless of the method we use, we eventually come to Equation (5.27).

Observe that from (5.27) it follows that:

(i) If $D=0$, then $Ax+By+Cz=0$ represents a plane passing through the origin;
(ii) If $C=0$, then $Ax+By+D=0$ represents a plane parallel to the Z-axis;
(iii) If $B=0$, then $Ax+Cz+D=0$ represents a plane parallel to the Y-axis; and
(iv) If $A=0$, then $By+Cz+D=0$ represents a plane parallel to the X-axis.

Example 5.5.6
What condition has to be satisfied in order for the plane $Ax+By+Cz+D=0$

(i) to be parallel to the YZ-plane?
(ii) to have equal intercepts on the Y- and Z-axes?

Solution

(i) $B=C=0$.
(ii) $B=C$. ■

Example 5.5.7
Find the equation of the plane determined by the point $P(1,1,1)$ and the vectors $\boldsymbol{a}=(1,-1,1)$ and $\boldsymbol{b}=(2,3,-1)$.

Solution

$$\begin{vmatrix} x-1 & y-1 & z-1 \\ 1 & -1 & 1 \\ 2 & 3 & -1 \end{vmatrix} = -2x+3y+5z-6=0 \quad ■$$

Example 5.5.8
Find the equation of the plane Π in intercept form that passes through the points $P_1(1,1,0), P_2(1,0,1)$ and $P_3(0,1,1)$.

Solution
Since points P_1, P_2, and P_3 lie in the plane Π, the vectors $\overrightarrow{P_1P_2}=(0,-1,1)$ and $\overrightarrow{P_1P_3}=(-1,0,1)$ are also in Π. Hence, the equation of the plane Π is

$$\begin{vmatrix} x-1 & y-1 & z \\ 0 & -1 & 1 \\ -1 & 0 & 1 \end{vmatrix} = -x-y-z+2=0$$

i.e.

$$-x-y-z=-2$$

and therefore

$$\frac{x}{2}+\frac{y}{2}+\frac{z}{2}=1.$$

■

Example 5.5.9
Find the equation of the plane containing the vector $\boldsymbol{a}=\boldsymbol{k}$ and two points P_1 and P_2 whose position vectors are $\boldsymbol{r}_1=-\boldsymbol{i}+2\boldsymbol{j}+3\boldsymbol{k}$ and $\boldsymbol{r}_2=3\boldsymbol{i}+\boldsymbol{j}+4\boldsymbol{k}$.

Solution

$$\boldsymbol{r}=\boldsymbol{r}_2-\boldsymbol{r}_1=4\boldsymbol{i}-\boldsymbol{j}+\boldsymbol{k}.$$

Hence, the equation of the plane is

$$\begin{vmatrix} x+1 & y-2 & z-3 \\ 4 & -1 & 1 \\ 0 & 0 & 1 \end{vmatrix}=-x-4y+7=0.$$

■

Exercise 5.5.4
Let Π be a plane passing through a point $P(x_0,y_0,z_0)$ and perpendicular to the vector $\boldsymbol{n}=(A,B,C)$. Show that the equation of Π is

$$A(x-x_0)+B(y-y_0)+C(z-z_0)=0.$$

Let's look once again at the distance between a point and a plane.

Example 5.5.10
Find the distance δ between the point $P(1,-2,-4)$ and the plane $\Pi: 2x+2y-z-11=0$ (Figure 5.15).

Solution
If we choose any point in the plane, say, $P_0(3,2,-1)$, (indeed: $2\cdot 3+2\cdot 2-(-1)-11=0$), then

$$\overrightarrow{P_0P}=(1-3)\boldsymbol{i}+(-2-2)\boldsymbol{j}+(-4-(-1))\boldsymbol{k}$$
$$=-2\boldsymbol{i}-4\boldsymbol{j}-3\boldsymbol{k}.$$

On the other hand, a unit vector perpendicular to Π is

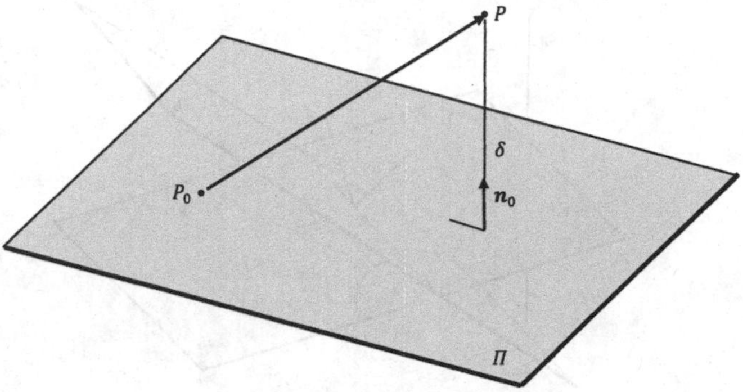

FIGURE 5.15

$$n_0 = \frac{2}{3}i + \frac{2}{3}j - \frac{1}{3}k.$$

Therefore

$$\delta = \left|\overrightarrow{P_0P} \cdot n_0\right| = \left|(-2i - 4j - 3k) \cdot \left(\frac{2}{3}i + \frac{2}{3}j - \frac{1}{3}k\right)\right|$$

$$= \left|-\frac{4}{3} - \frac{8}{3} + \frac{3}{3}\right| = 3.$$

■

5.6 The Angle Between Two Planes

Definition 5.6.1

Let Π_1 and Π_2 be two intersecting planes in $\mathbf{R}^3$. By the angle $\phi = \measuredangle(\Pi_1, \Pi_2)$ formed by Π_1 and Π_2 we mean the smaller of the two supplementary angles between them, i.e. $\leq \frac{\pi}{2}$ (Figure 5.16). The size of that angle is the same as that of the angle between normal vectors n_1 and n_2 to the respective planes.

So, if Π_1 and Π_2 are given by

$$\Pi_1 : A_1x + B_1y + C_1z + D_1 = 0,$$

$$\Pi_2 : A_2x + B_2y + C_2z + D_2 = 0,$$

the corresponding normal vectors are $n_1 = (A_1, B_1, C_1)$ and $n_2 = (A_2, B_2, C_2)$. So

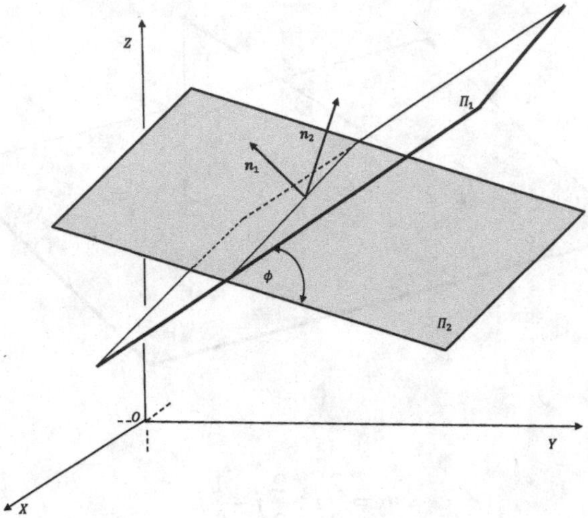

FIGURE 5.16

$$\cos\phi = \cos(\Pi_1, \Pi_2) = \cos(n_1, n_2)$$

$$= \frac{|n_1 \cdot n_2|}{n_1 \cdot n_2} = \frac{|A_1A_2 + B_1B_2 + C_1C_2|}{\sqrt{A_1^2 + B_1^2 + C_1^2}\sqrt{A_2^2 + B_2^2 + C_2^2}}.$$

Evidently, planes Π_1 and Π_2 are perpendicular iff

$$A_1A_2 + B_1B_2 + C_1C_2 = 0.$$

Likewise, planes Π_1 and Π_2 are parallel if vectors normal to the planes are parallel, i.e.

$$n_1 = \lambda n_2.$$

This condition may be expressed as

$$\frac{A_1}{A_2} = \frac{B_1}{B_2} = \frac{C_1}{C_2} \neq \frac{D_1}{D_2}.$$

In other words, if the system of two linear equations

$$\left.\begin{aligned} A_1x + B_1y + C_1z + D_1 = 0 \\ A_2x + B_2y + C_2z + D_2 = 0 \end{aligned}\right\} \tag{5.28}$$

representing planes Π_1 and Π_2, is inconsistent, then the planes are parallel and the coordinates of the normal vectors $n_1 = (A_1, B_1, C_1)$ and

$n_2 = (A_2, B_2, C_2)$ are proportional. Conversely, if the system (5.28) is consistent and the equations are proportional, i.e.

$$\frac{A_1}{A_2} = \frac{B_1}{B_2} = \frac{C_1}{C_2} = \frac{D_1}{D_2},$$

then $\Pi_1 = \Pi_2$.

Example 5.6.1
Find the angle ϕ between the planes

$$\Pi_1 : x + 2y + 2z - 10 = 0 \text{ and } \Pi_2 : x + y + 4z - 7 = 0.$$

Solution

$$\cos\phi = \cos(n_1, n_2) = \frac{(-1)\cdot 1 + 2\cdot 1 + 2\cdot 4}{\sqrt{(-1)^2 + 2^2 + 2^2}\sqrt{1^2 + 1^2 + 4^2}} = \frac{1}{\sqrt{2}}.$$

Hence, $\phi = 45°$. ■

Example 5.6.2
Show that the planes
$\Pi_1 : 2x - y + 2z - 3 = 0$ and $\Pi_2 : 2x + 2y - z - 7 = 0$
are perpendicular.

Solution
Observe that the vectors $n_1 = 2i - j + 2k$ and $n_2 = 2i + 2j - k$ are perpendicular to the respective planes. On the other hand,

$$n_1 \cdot n_2 = (2i - j + 2k)\cdot(2i + 2j - k) = 0$$

Thus, planes Π_1 and Π_2 are perpendicular. ■

Exercise 5.6.1
Prove that:

(i) $\Pi_1 : x + 3y + z + 4 = 0$ and $\Pi_2 : 2x - y + z + 2 = 0$,
(ii) $\Pi_1 : 3x - 2y + z + 7 = 0$ and $\Pi_2 : 9x - 6y + 3z + 10 = 0$,

are parallel.

Finally, notice that the line l determined by two intersecting planes (i.e., common to both planes) is obtained in the following way:

Consider two planes

$$\left.\begin{aligned} \Pi_1 &: \quad \boldsymbol{n}_1 \cdot \boldsymbol{r} + D_1 = 0 \\ \Pi_2 &: \quad \boldsymbol{n}_2 \cdot \boldsymbol{r} + D_2 = 0 \end{aligned}\right\} \tag{5.28}$$

A point P is on the line l formed by two intersecting planes iff the position vector $\boldsymbol{r}$ specifies a point lying in both planes (Figure 5.17).

So, let's repeat: since an equation of the form

$$Ax + By + Cz + D = 0$$

represents a plane, two such equations considered simultaneously represent two planes which, if not parallel, intersect in a straight line.

Of course, many planes may pass through the same line (in fact infinitely many), and therefore a line may be determined by any one of infinitely many pairs of planes through the line. The set of all planes through a line is called a ***sheaf of planes***, (Figure 5.18).

Example 5.6.3
Let

$$\Pi_1 : A_1 x + B_1 y + C_1 z + D_1 = 0,$$

$$\Pi_2 : A_2 x + B_2 y + C_2 z + D_2 = 0,$$

be two intersecting planes. Find the *direction numbers*, i.e. the components, of the vector of the line common to Π_1 and Π_2.

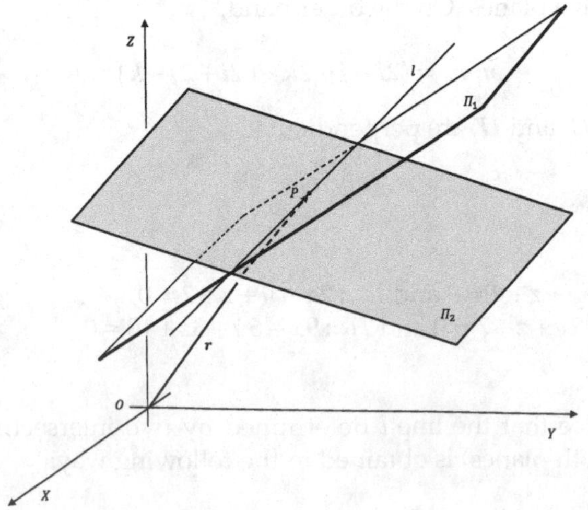

FIGURE 5.17

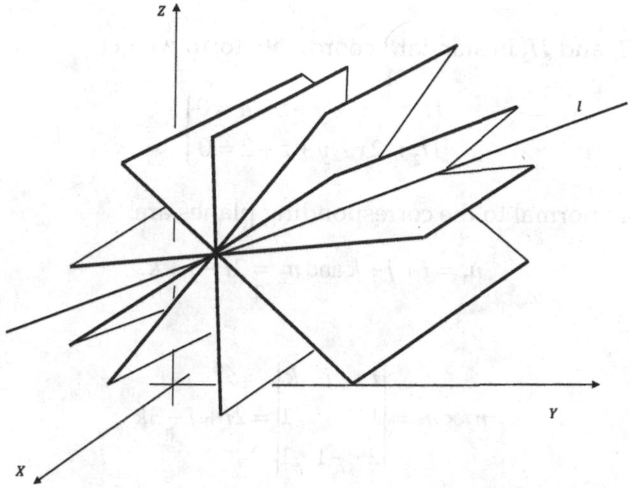

FIGURE 5.18

Solution
The following vectors are normal to Π_1 and Π_2, respectively:

$$\boldsymbol{n}_1 = A_1\boldsymbol{i} + B_1\boldsymbol{j} + C_1\boldsymbol{k}$$

and

$$\boldsymbol{n}_2 = A_2\boldsymbol{i} + B_2\boldsymbol{j} + C_2\boldsymbol{k}$$

The line determined by the two planes is perpendicular to the vectors normal to the planes, i.e. it is parallel to

$$\boldsymbol{n}_1 \times \boldsymbol{n}_2 = \begin{vmatrix} \boldsymbol{i} & \boldsymbol{j} & \boldsymbol{k} \\ A_1 & B_1 & C_1 \\ A_2 & B_2 & C_2 \end{vmatrix} = (B_1C_2 - B_2C_1)\boldsymbol{i} - (A_1C_2 - A_2C_1)\boldsymbol{j} + (A_1B_2 - A_2B_1)\boldsymbol{k}.$$

Hence, the direction numbers we are looking for are:
$(B_1C_2 - B_2C_1), (A_1C_2 - A_2C_1)$ and $(A_1B_2 - A_2B_1)$. ■

Example 5.6.
Find the equation of the line l common to the planes:

$$\Pi_1 : \quad (\boldsymbol{i} + \boldsymbol{j} + \boldsymbol{k})\cdot\boldsymbol{r} - 4 = 0$$

$$\Pi_2 : \quad (2\boldsymbol{i} - \boldsymbol{j} + \boldsymbol{k})\cdot\boldsymbol{r} + 2 = 0$$

Solution
Rewriting Π_1 and Π_2 in standard coordinate form we get

$$\left.\begin{aligned} \Pi_1 :&\quad x+y+z-4=0 \\ \Pi_2 :&\quad 2x-y+z+2=0 \end{aligned}\right\} \tag{5.30}$$

Thus, vectors normal to the corresponding planes are:

$$\boldsymbol{n}_1 = \boldsymbol{i}+\boldsymbol{j}+\boldsymbol{k} \text{ and } \boldsymbol{n}_2 = 2\boldsymbol{i}-\boldsymbol{j}+\boldsymbol{k}.$$

So

$$\boldsymbol{n}_1 \times \boldsymbol{n}_2 = \begin{vmatrix} \boldsymbol{i} & \boldsymbol{j} & \boldsymbol{k} \\ 1 & 1 & 1 \\ 2 & -1 & 1 \end{vmatrix} = 2\boldsymbol{i}+\boldsymbol{j}-3\boldsymbol{k} \tag{5.31}$$

Now, if $x = 0$, system (5.29) becomes

$$\left.\begin{aligned} y+z-4 &= 0 \\ -y+z+2 &= 0 \end{aligned}\right\} \tag{5.32}$$

Solving (5.31) we get $y = 3$ and $z = 1$, so the point $P(0,3,1)$ is on the line we are looking for. Finally, with (2) and the coordinates of the point, we can write the canonical equation of the line l

$$\frac{x}{2} = \frac{y-3}{1} = \frac{z-1}{-3}$$

Equivalently, we can write

$$\boldsymbol{r} = 3\boldsymbol{j}+\boldsymbol{k}+\lambda(2\boldsymbol{i}+\boldsymbol{j}-3\boldsymbol{k}).$$ ■

Another, equivalent, approach is illustrated in the following example.

Example 5.6.5
Find the equation of the line l common to the planes

$$\left.\begin{aligned} \Pi_1 :&\quad 3x-y+z-8=0 \\ \Pi_2 :&\quad 2x+y+4z-2=0 \end{aligned}\right\} \tag{5.33}$$

Solution
Let's first find the two points (the *piercing points*) where the line l passes through two of the coordinate planes, the XY-plane and the YZ-plane. If we take $z = 0$ then from (5.33) we get

$$\left.\begin{aligned} 3x-y-8&=0\\ 2x+y-2&=0 \end{aligned}\right\} \tag{5.34}$$

Solving (5.34) we get one piercing point $A(2,-2,0)$ in the XY-plane. Similarly, if $x=0$ then from (5.33) we get

$$\left.\begin{aligned} -y+z-8&=0\\ y+4z-2&=0 \end{aligned}\right\} \tag{5.35}$$

Solving (5.35) we get the piercing point $B(0,-6,2)$ in the YZ-plane. Now it is easy to construct a vector $\overrightarrow{AB}=-2i-4j+2k$ which gives us direction numbers for the line $AB=l$. Hence, the canonical equation of l is

$$\frac{x-2}{-2}=\frac{y+2}{-4}=\frac{z}{2}.$$

■

Note

1 Ludwig Otto Hesse (1811–1874), German mathematician.

Appendix A

A.1 Sets

The word *set*, introduced in 1897 by Georg Cantor,[1] is used to indicate a collection, "collection into a whole", of definite and separate objects, i.e., objects that have some (at least one) property that separates them from everything else. These objects are called the elements or members of the set.

If A is a set, the notation $a \in A$ means that a is an element of A, and $x \notin A$ means that x is not a member of A. This is graphically represented by a *Venn diagram* (Figure A.1).

To list the elements of a set explicitly, we write

$$A = \{a,\ b,\ c,\ d,\ e,\ f\}$$

meaning that $a \in A, b \in A, c \in A$, etc.

If A is a set and $P(x)$ is a property that is satisfied by all elements of A, we write

$$A = \{x \mid P(x)\}$$

which is read as "A is the set of all x's such that $P(x)$", i.e., all those elements x satisfying property P. More formally, we have

Axiom of Extensionality: *A set is uniquely determined by the elements it contains, i.e. two sets are considered equal if they have the same elements. This is often, but less clearly, stated as: A set is determined by its extension. We write*

$$A = B \leftrightarrow (\forall x)(x \in A \leftrightarrow x \in B).$$

Definition A.1

Given two sets A and B, we say that B is a subset of A, and we write $B \subseteq A$, if and only if every element of A is also an element of B, i.e.

$$B \subseteq A \leftrightarrow (\forall\ x \in B,\ x \in A),$$

(Figure A.2).

Definition A.2

Given two sets A and B we define the **union** of A and B, denoted by $A \cup B$, to be the set of all elements that are in at least one of A or B, i.e.,

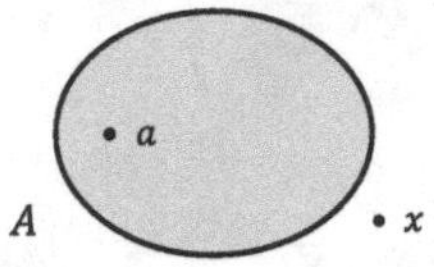

FIGURE A.1

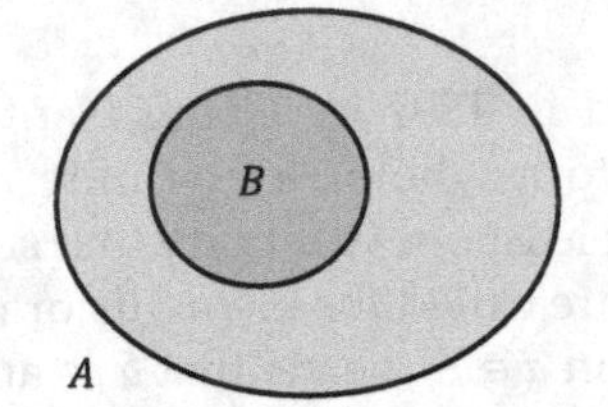

FIGURE A.2

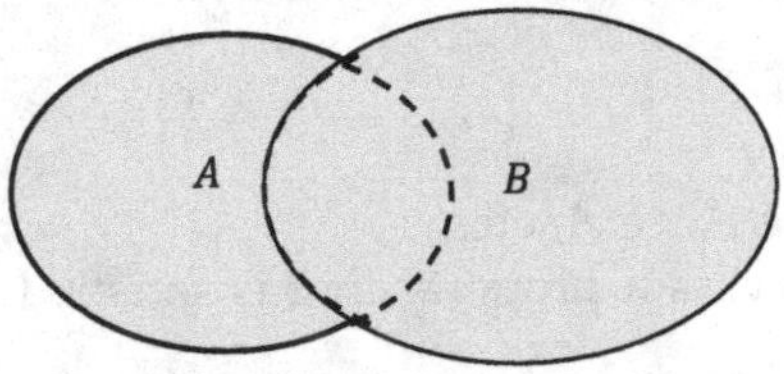

FIGURE A.3

$$A \cup B = \{x \mid x \in A \text{ or } x \in B\},$$

(Figure A.3).

The union of a collection of n sets is defined analogously:

$$\bigcup_{i=1}^{n} A_i = A_1 \cup A_2 \ldots \cup A_n = \{\, x \mid x \in A_i \text{ for at least one } i = 1, \ldots n \,\}.$$

Definition A.3

Let A and B be sets. The **intersection** of A and B, denoted by $A \cap B$, is the set of all elements that are common to A and B, i.e.

$$A \cap B = \{x \mid x \in A \;\&\; x \in B\},$$

(Figure A.4).

The intersection of a collection of n sets is defined analogously:

$$\bigcap_{i=1}^{n} A_i = A_1 \cap A_2 \cap \ldots \cap A_n = \{\, x \mid x \in A_i, \text{ for all } i = 1, 2, \ldots n \,\}.$$

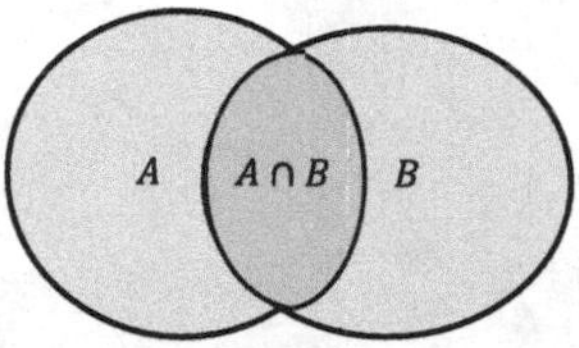

FIGURE A.4

Note

1 Georg Ferdinand Ludwig Philipp Cantor (1845–1918), German mathematician.

Appendix B

B.1 Sets of Numbers

Natural numbers:

$$\mathbf{N} = \{1,2,3,\ldots\}$$[1]

Integers:

$$\mathbf{Z} = \{\ldots -3,-2,\ -1,0,1,2,3,\ldots\};$$

Rational numbers:

$$\mathbf{Q} = \left\{ x \mid x = \frac{p}{q},\ p,q \in \mathbf{Z}\ ,q \neq 0 \right\}.$$

Notice that $\mathbf{N} \subseteq \mathbf{Z} \subseteq \mathbf{Q} \subseteq \mathbf{R}$ (Figure B.1).

B.2 Properties of the Real Numbers

Consider the set **R**, with two binary operations, "+", called *addition*, and "·", called *multiplication*, then we want the following to hold:

$\mathcal{A}$ **1.1** $\forall a,b \in \mathbf{R},\ a+b=b+a \in \mathbf{R};$
$\mathcal{A}$ **1.2** $\forall a,b \in \mathbf{R},\ a\cdot b=b\cdot a \in \mathbf{R};$
$\mathcal{A}$ **2.1** $\forall a,b,c \in \mathbf{R},\ (a+b)+c=a+(b+c);$
$\mathcal{A}$ **2.2** $\forall a,b,c \in \mathbf{R},\ (a\cdot b)\cdot c=a\cdot(b\cdot c);$
$\mathcal{A}$ **3.1** $\forall a,b,c \in \mathbf{R},\ a\cdot(b+c)=a\cdot b+a\cdot c;$
$\mathcal{A}$ **3.2** $\forall a,b,c \in \mathbf{R},\ (b+c)\cdot a=b\cdot a+c\cdot a;$
$\mathcal{A}$ **4.1** $\exists 0 \in \mathbf{R}, 0+a=a+0=a,\ \forall a \in \mathbf{R};$
$\mathcal{A}$ **4.2** $\exists 1 \in \mathbf{R}, 1\cdot a=a\cdot 1=a,\ \forall\ a \in \mathbf{R};$
$\mathcal{A}$ **5.1** $\forall\, a \in \mathbf{R},\ \exists (-a) \in \mathbf{R},\ a+(-a)=(-a)+a=0;$
$\mathcal{A}$ **5.2** $\forall\, a \in \mathbf{R}$ if $a \neq 0, \exists\, a^{-1} \in \mathbf{R}, a\cdot a^{-1} = a^{-1}\cdot a = 1.$

In mathematics, a **field** Φ is a set on which two binary operations, denoted "+" and "·", called "addition" and "multiplication", are defined such that the

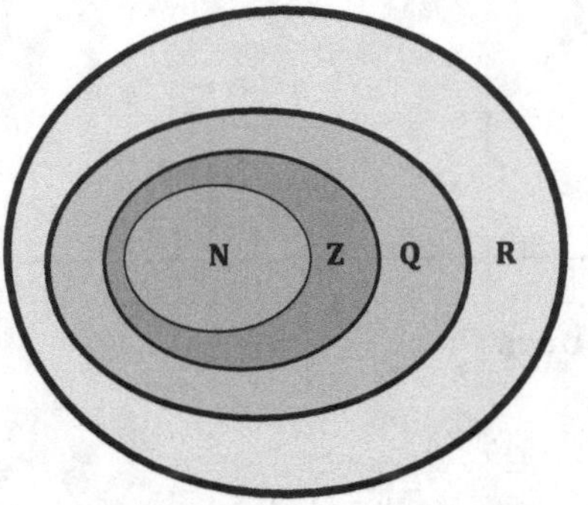

FIGURE B.1

corresponding *inverse* operations "subtraction" and "division" also exist, i.e., such that the set is closed with respect to these operations. These operations have to satisfy the following axioms:

$F1.1\ \forall \alpha,\beta \in \Phi,\ \alpha+\beta=\beta+\alpha \in \Phi;$

$F1.2\ \forall \alpha,\beta \in \Phi, \alpha\cdot\beta=\beta\cdot\alpha \in \Phi;$

$F2.1\ \exists 0 \in \Phi,\ \alpha+0=0+\alpha=\alpha,\ \forall \alpha \in \Phi;$

$F2.2\ \exists 1 \in \Phi,\ \alpha\cdot 1=1\cdot\alpha=\alpha,\ \forall \alpha \in \Phi;$

$F3.1\ \forall\ \alpha \in \Phi, \exists(-\alpha) \in \Phi,\ \alpha+(-\alpha)=(-\alpha)+\alpha=0;$

$F3.2\ \forall\ \alpha \neq 0 \in \Phi, \exists\ \alpha^{-1} \in \Phi,\ \alpha\cdot\alpha^{-1}=\alpha^{-1}\cdot\alpha=1;$

$F4.1\ \forall \alpha,\beta,\gamma \in \Phi,\ \alpha+(\beta+\gamma)=(\alpha+\beta)+\gamma;$

$F4.2\ \forall \alpha,\beta,\gamma \in \Phi,\ \alpha\cdot(\beta\cdot\gamma)=(\alpha\cdot\beta)\cdot\gamma;$

$F5.1\ \forall \alpha,\beta,\gamma \in \Phi,\ \alpha\cdot(\beta+\gamma)=(\alpha\cdot\beta)+(\alpha\cdot\gamma);$

$F5.2\forall \alpha,\beta,\gamma \in \Phi,\ (\beta+\gamma)\cdot\alpha=(\beta\cdot\alpha)+(\gamma\cdot\alpha).$

It follows that the set **R,** satisfying axioms **$\mathcal{A}$ 1.1 – A 5.2**, is a field.

If $a,b,c \in \mathbf{R}$ then the following also hold:

T.1 If $a+b=a+c$, then $b=c$;
T.2 $a-b=a+(-b)$;
T.3 $-(-a)=a$;
T.4 $(-a)\cdot b=a\cdot(-b)=-(a\cdot b)$;
T.5 $a\cdot(b-c)=a\cdot b-a\cdot c$;
T.6 $0\cdot a=a\cdot 0=0$;
T.7 If $a\cdot b=0$, then $a=0$ or $b=0$;

T.8 If $a \cdot b = a \cdot c$ and $a \neq 0$, then $b = c$;
T.9 If $b \neq 0$, then $a / b = a \cdot b^{-1}$;
T.10 If $a \neq 0$, then $\left(a^{-1}\right)^{-1} = a$;
T.11 Exactly one of the following is true: $a\langle b,a\rangle b$, or $a = b$;
T.12 If $a < b$, then $a + c < b + c$;
T.13 If $a < b$ and $c > 0$, then $ac < bc$;
T.14 If $a < b$, then $-a > -b$;

Note

1 Some authors include 0 in **N**.

Appendix C

With each square matrix A, i.e., a rectangular array of scalars usually represented as

$$A = \begin{bmatrix} a_{11} & a_{12} & \cdots & a_{1n} \\ a_{21} & a_{22} & \cdots & a_{2n} \\ \vdots & \vdots & \cdots & \vdots \\ a_{n1} & a_{n2} & \cdots & a_{nn} \end{bmatrix},$$

we can associate a special scalar called the **determinant** of A, denoted by $\det A$, represented as

$$\det A = \begin{vmatrix} a_{11} & a_{12} & \cdots & a_{1n} \\ a_{21} & a_{22} & \cdots & a_{2n} \\ \vdots & \vdots & \cdots & \vdots \\ a_{n1} & a_{n2} & \cdots & a_{nn} \end{vmatrix}.$$

Determinants of order 1, 2, and 3 are defined as follows:

$$|a_{11}| = a_{11},$$

$$\begin{vmatrix} a_{11} & a_{12} \\ a_{21} & a_{22} \end{vmatrix} = a_{11}a_{22} - a_{12}a_{21},$$

$$\begin{vmatrix} a_{11} & a_{12} & a_{13} \\ a_{21} & a_{22} & a_{23} \\ a_{31} & a_{32} & a_{33} \end{vmatrix} = a_{11}(a_{22}a_{33} - a_{23}a_{32}) - a_{12}(a_{21}a_{33} - a_{23}a_{31}) + a_{13}(a_{21}a_{32} - a_{22}a_{31}).$$

Index

Printed in the United States
by Baker & Taylor Publisher Services

Printed in the United States
by Baker & Taylor Publisher Services